CREATING·A
WILDLIFE
GARDEN

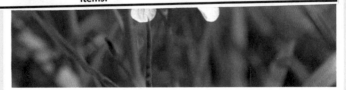

Introduction

The last few decades have seen drastic changes in the countryside of northern Europe – changes which have caused a dramatic decrease in the range and abundance of our native wildlife. At the same time, there has been a tremendous upsurge of interest in our natural heritage and its conservation, as this becomes increasingly urgent. At one time, not so very long ago, gardening could not have seemed further removed from the ideas of nature conservation, in an era when control and tidiness were important, pesticides were all-pervading, and the ideas of ecological management and re-creation of habitats were in their infancy. While much of this may still be true, there has been a steady growth of the idea that gardens can, and should, play a part in this natural movement – oases of diverse, semi-natural, and controlled environments, where nature can find some sanctuary from pesticides and habitat loss. With so much of our land under housing development, the area of gardens increases year by year at the expense of countryside, and so its relative importance grows.

At first, wildlife gardening generally consisted of putting a few extras into the ordinary garden, mainly for birds – food in winter, nestboxes, and so on. This has gradually increased to include provision for other creatures, especially mammals and insects, using nectar plants, hedgehog homes, bat boxes and other features. However, just as the emphasis in nature conservation in general, and nature reserve management, has shifted towards managing habitats rather than species, so the emphasis in wildlife gardens has changed. It stems from a realisation that all living creatures are inter-related in some way or other, and there is little point in managing for a few rare species when everything around them is changing. What point is there in batboxes for bats to roost in, if there are not enough insects for the bats to feed on to survive? What butterflies will there be to come to a Buddleia bush if those butterflies cannot find the right plants to lay their eggs on? The more we discover about the natural world, the more complex it proves to be, and the most logical response to this is to manage or re-create habitats that we know to be successful in providing homes for a wide variety of plants and animals.

In gardens, this has spawned the idea of recreating miniature examples of natural habitats, often in combinations that would occur naturally, although of course on a much smaller scale and at a much higher density. The proof of this particular pudding has been in the eating – the relatively few people (with a clear knowledge of both their subject and their aims) who have pioneered the ideas and the techniques, have shown by their results that such garden mini-habitats can indeed be very rich in species. It is true that there will always be a great number of species, especially among mammals and birds, that will never *directly* benefit from gardens, however successfully they are managed for wildlife; but nevertheless, there is a vast range of species whose abundance would increase if more gardens were managed in this way. Again, the influence of a successful wildlife garden probably extends much further than you imagine, for a heathland hobby might easily prey on the dragonflies from your pond or the house martins raised in a nest under your eaves, and a rare species of bat might feed on moths nurtured in your garden! It really does work, if you get it right, and you will find that, not only do you attract a lot more wildlife to your garden, but that much more of it will take up actual residence there.

The most recent step in the current evolutionary process of wildlife gardening is the idea of actually *designing* the wildlife garden – in other words, giving a great deal of preliminary thought to what you have, what you want, and how you can create a wildlife garden in a particularly attractive way. The increased awareness that comes from this preparatory work can enable you, not only to create the optimum conditions for a more successful wildlife garden, but to make a beautiful and varied garden that fulfils the needs of *all* its users.

This is really the subject of this book – how to design and create a garden that will be attractive and rich in wildlife, but also satisfying, productive and above all, absolutely fascinating and enjoyable. We hope it stimulates at least a few people to achieve this aim.

CREATING · A
WILDLIFE
GARDEN

BOB AND LIZ GIBBONS

HAMLYN

First published in 1988
by The Hamlyn Publishing Group Limited,
a Division of The Octopus Publishing Group plc.,
Michelin House, 81 Fulham Road, London SW3 6RB

ISBN 0 600 33384 1

Printed in Spain

CONTENTS

WILDLIFE BY DESIGN

BRINGING WILDLIFE INTO YOUR GARDEN

CREATING NEW GARDEN HABITATS

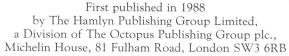

Nature of the garden

Gardens are very important places for wildlife. No doubt they always have been, but today their importance has increased immensely as bastions against the changes that have gone on all around them. For the damage wrought in our countryside over the last 30 years or so have been quite devastating to natural and near-natural habitats and the associated wildlife; yet, gardens have changed relatively little. With a little thought and planning, and even sometimes by fortunate accident, gardens can become home to hundreds of creatures of all kinds, from hoverflies to hedgehogs. They can never *replace* genuinely wild habitats, and there are many species, especially among the larger mammals and birds, which will never make any use of a garden however large or well-designed, but for a considerable number they are becoming, literally, a life-saver.

The garden environment is quite unlike any normal piece of countryside in many ways, though the ordinary ecological principles still apply. Paradoxically, the more you apply the designs and ideas outlined in this book, the less like the countryside your garden will become because no piece of natural habitat is as diverse as a well-designed wildlife garden. It is this diversity

As more and more natural countryside is destroyed by agriculture and development, so wildlife gardens play a more and more important role.

that is the key both to its success in attracting wildlife, and its failures. For, the more you split up your garden into little areas of this and that, the smaller each one will be. For example, if you were lucky enough to have an acre of garden and you decided to devote it entirely to coppiced hazel with standard oaks, then you might expect to have a reasonable proportion of the coppice specialities such as nightingales or dormice eventually moving in; if, however, you divided up the garden into a series of 30 or so different habitat sections, with coppice represented by just a few hazel bushes, then you would probably end up with many more common species overall, but fewer of the specialist plants or

These flowers, in an organic garden in Germany, are an irresistible attraction to insect life, and their careful selection for colour makes a special feature.

animals. This dilemma of size versus diversity is an ever-present one, but most people will opt for the pleasure of having the mass of species that diversity brings, especially as few of us own enough land to create or maintain a really worthwhile area of one habitat. We will come back to this dilemma in the chapters on designing the wildlife garden.

Variety in the garden

You might, at first, think that your own garden is not very varied, whether it is a wildlife garden or not. But, if you step outside the door and look at it in detail with natural history in mind, you begin to realise just how diverse even the most ordinary garden is.

Just outside the back door there is probably an area of paving or concrete. This is hardly outstanding for wildlife, but any cracks between the stones will serve as homes for plants like hairy bitter-cress, for ants' nests, for ground beetles or spiders. It is an area where birds will readily find any food that you put out, and its very bareness may help them to feed with a clear view of potential predators.

13

Paving does not have to be tidily maintained! The cracks between these old tiles are filled with bluebells, making an attractive display.

Next to this, there may be a fence or wall. If it is well maintained and regularly creosoted, then its value will be limited. (As a general rule, creosote is not to be recommended as it can be toxic to all forms of life, including plants growing next to the fence, so it is better to use one of the more modern alternatives.) If the fence is fairly neglected, it will be used by butterflies and moths as pupating sites; lichens, and algae mosses will grow on it, and even some small flowering plants may take root; these in turn will be fed upon by a variety of invertebrates. Birds will use any fence or wall as a song or lookout perch. If your boundary structure is covered by plants, then it takes on a whole new dimension.

Close by, there may be a shed, which will share some of the qualities of a fence, but with the added advantage of having a covered area beneath. If you were able simply to lift up your shed and look underneath, you would probably be amazed at what was there — hedgehog nests, bumble bee nests, slow worms, voles and a mass of smaller invertebrates live in such places, though most only emerge at night, if at all, so you rarely know that they are there.

Somewhere, there is almost bound to be a lawn, at least in most British gardens though not in all other countries. Lawns vary enormously in their interest and value, from the regularly-fertilised, weed-free, tightly-cut, bright green version through to the rough flowery type that borders on being a meadow. All of them serve as feeding areas for birds that come to take the vast invertebrate populations which live among the roots, but they can offer much more than this, if allowed a little tolerance, as discussed from page 112.

It is the flower and shrub borders that are the most noticeably attractive parts of the garden, and these are, perhaps, the most unreal of all. By a combination of cultivation, careful nurturing through the young stages, weeding, and a thoughtful selection process, we persuade dozens, maybe hundreds, of different plants, from all parts of the world, to grow together in our herbaceous or shrub borders, providing many opportunities for insects, birds and other forms of wildlife. But, as we shall see later, the choice of species you plant can have a considerable influence on the wildlife you attract, and this is one of the key features to consider in the making of a wildlife garden. The density and diversity of plants in these areas is somewhat unnatural, and we tend to make it even more so by choosing plants which will flower successively throughout the whole growing season, from spring bulbs in February or March through to the last flowers to be hit by the early winter frosts. In most natural situations, the period of intense flowering tends to be much shorter.

If you have a hedge (and every wildlife garden should have one), it will be broadly similar to those in the countryside, especially as it matures. You will, however, probably manage it much more selectively even if it is only a short length: for example, you may delay cutting until the hawthorn has flowered in May; carefully trim around the rose which flowers in June or leave the sloes in early autumn; you may allow it to thicken out in one area where it is convenient but trim it back elsewhere, and so on. The countryside hedges, by contrast, tend to be cut, flailed or laid in a uniform fashion once a year or less, so, once again, you are creating greater diversity in the garden by means of selective management.

Elsewhere in the garden, you probably have a tree or two supporting wildlife of all forms; perhaps some wooden posts for a washing line or part of a fruit cage which will offer homes to solitary wasps and other

invertebrates, or perching places for birds. A woodpile can house a surprising number of mammals, birds and insects, while even compost heaps have their own specialised flora and fauna, and they are one of the prime egg-laying sites for grass snakes! An old tree stump will provide food for wood-boring insects, such as stag beetles, and there may be bumble bees and small mammals living among the roots. Garages and outhouses offer similar nooks and crannies to the garden shed, but with the added dimension of rafters for nesting swallows, uprights for wrens and the old door hinge for spotted flycatchers!

If you have a pond, a whole new dimension is added and the possibilities for extra species are enormous (see page 128 onwards). Once again, it will probably be more diverse than a 'wild' pond of similar size because you will stock it with plants, clear out the fallen leaves, remove the blanketing algae periodically and generally watch over its state of health.

Finally, of course, there is the house itself which adds the possibility of swifts in the loft, house martins under the eaves, bats in the roof space or under the hanging tiles – the list goes on.

So, without mentioning all the other possible features, from gravelled areas to rock gardens, it is

---

Bare walls are of limited value for wildlife, but well-covered walls, such as this one draped with Senecio greyii, *can be full of life.*

Birds will nest in all sorts of odd places, provided they are not disturbed. This spotted flycatcher has made a home on an old door hinge.

immediately obvious that the garden is an incredibly varied and diverse mosaic, more so than almost any comparable area of countryside. And, the way in which you design and fill this mosaic can have a tremendous effect on the wildlife that you attract to it.

To give an idea of what this diversity means in terms of numbers of species, you can reckon that in an average, medium-sized garden, managed sympathetically for wildlife, you might expect:

- *at least 250 species or varieties of flowering plants;*
- *50 or 60 species of birds, resident or visiting;*
- *15 or so butterfly species;*
- *five or six different dragonfly species;*
- *hundreds, or even thousands of insect and other invertebrate species;*
- *together with occasional reptiles, amphibians, mammals and lower plants such as mosses or lichens.*

The garden jigsaw

You may view your garden as a series of separate entities such as lawn, borders, paths and compost heap, but, for the vast majority of animals, it is the sum of the parts that is more important than the whole. With the exception of very small or extremely immobile species, virtually all forms of animal life from aphids to foxes, use more than one ecological 'niche'. For some, your

GARDEN DESIGN PLANS

The design plans in this book are all based on actual gardens, and will differ in many respects from that of the reader. This should not matter in itself, as the ideas can be adapted to virtually any garden. We hope that, with the aid of this book, you will be able to work out a design tailor-made for your garden. It will fit both the site – in terms of its physical character and of the framework of line, shape, and volume that you create – and your requirements – in terms of what you then put into that basic framework to meet your needs and those of the wildlife.

The designs do not include detailed planting plans, as the aim is to demonstrate how different *types* of planting are useful in particular situations. For example, a block of planting will be defined as 'medium-height species-rich mixed hedgerow with native climbers', and not by the individual species which it might contain. The latter will vary with many factors – including personal preference, as there are very few instances where there are not equally valid alternatives. We would advise the reader to consult the lists of suggested plants in this, and perhaps other recommended books, before making a final choice.

WILDLIFE FEATURES IN A MIXED-HABITAT GARDEN

This informal garden contains a large number of differing habitats used as elements of an integrated design. The garden is a general one in many ways, as it is both a general wildlife garden, and also a garden in which family activities can be accommodated. A point well demonstrated by this design is that a functional balance is maintained in the garden if there is adequate provision of space for everyday activities – patio, mown lawn, play area and so on.

The design uses a flowing line to define one large space – **lawns, meadow, and wetland** – bounded by various smaller habitats, which shelter and enclose the rest of the garden, and are valuable to wildlife in themselves – **nectar border, cornfield edge, shrub borders, and woodland edge.** The garden becomes progessively wilder away from the house, but paths allow access to all parts of the garden.

Around the house –
Wildlife will find a home here anyway, but you can still increase provision for it.
A *Artificial nests under the eaves for house martins*
B *Attic untreated with harmful chemicals – bat entry-points left open*
C *Fences and walls used as bird perches, pupation sites and spider homes*
D *Thick climbers – cover for nesting birds and hibernating insects*

E *Paving joints and cracks – small plants, insects, other invertebrates*
Lawns and Flowerbeds –
Flowering shrubs and butterfly borders suit this 'tidy' part of the garden.
F *A safe place for birds to feed in winter*
G *Many culinary herbs attract bees, hoverflies and other insects*
H *Flowering shrubs give permanent structure to this area*

I Colourful herbaceous plants in the butterfly border

J Crannies in wall – pupation sites for butterflies, spider homes etc

K Many garden birds find food on a traditional mown lawn

The Garden Pond – Paving gives visual definition to the division between pond and meadow, and access in all seasons.

L Amphibians hibernate in gaps under un-mortared slabs

M Insects and amphibians in the wetland fringe – a complement to the pond

N A pond is a bonus for wildlife – the hub of a good wildlife garden

O Unpointed paving joints for invertebrate habitat

The Cornfield Edge – Though re-sown yearly, this is a well-defined area which is an integral part of the design

P Flowery hedge gives food, breeding sites, and shelter for nearby areas

Q Once-common arable weeds are attractive and ecologically valuable

The Meadow – Routine mowing determines the patterns of different areas, and allows access without disturbance.

R The flowery lawn – good for ground-breeding insects – low-growing plants give nectar for foraging bees

S Thick hedge gives cover for birds and shelters the meadow, making it more attractive to butterflies

T Tall flowery meadow-grass – perfect for small mammals and butterflies

The Woodland – In one sense, this is the anchor point of the whole garden – the solid backdrop against which the rest of the garden and its wildlife are seen.

U Native trees and shrubs bring an extra dimension to the garden, attracting in wildlife from outside its boundaries

V Nest-boxes and bat-boxes in least-disturbed part of garden

W A clearing greatly increases the wildlife-value of the woodland

X Site log-piles for wood-boring insects, fungi, hedgehogs etc. in several places

Y Berry-bearing shrubs in a quiet part of the garden attract birds throughout winter

Z Garden shed – may be good nest-site for swallows if window left open

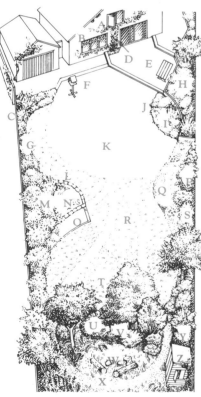

The pieces of the wildlife-garden jigsaw can be as small as a nestbox or as big as a meadow, and the picture endlessly variable.

garden will provide everything, while for others, such as badgers, it will only provide a small part of its extensive territory, and in extreme cases – such as migrant birds – your garden just forms one tiny stage in a range that may cover thousands of miles. On this larger scale, an individual garden becomes part of the urban or suburban network of green spaces, or part of the rural scene, depending upon your situation. But, on a smaller scale, the relationship between the different parts of your garden can be vital to its success or failure in attracting wildlife.

The society of garden life

If you consider, briefly, the life histories of one or two insects, it becomes clearer why this interconnection and juxtaposition of parts of a garden is so important.

A butterfly that frequently occurs in gardens is the beautiful small tortoiseshell. They are most often seen in gardens when large numbers arrive to visit ice plants *Sedum spectabile* in late summer, or on forms of *Buddleia* a little earlier in the year. But, as with many insects, these adult butterflies are just one part of a complicated life-cycle that has a number of stages. The insects that are seen in late summer are the second generation adults to emerge that year, and they will require late summer and autumn flowering nectar sources on which to feed to stock up with energy for the winter. At this stage, they do not breed, but simply

gather together in congenial sheltered flowery environments, like gardens, to prepare for hibernation. As winter approaches, these adults disperse to find hibernation sites (only a few butterfly species hibernate as adults, while most overwinter in some other form), and such sites are often associated with gardens. The most frequent places are in garages, open sheds, outhouses, woodstores and even in houses, with the main requirement being for somewhere cool and dry where they can remain hidden.

As soon as the weather becomes warmer in the following spring, the butterflies which have overwintered emerge. Their requirements now are for warm sheltered places where they can bask in the spring sunshine, some nectar-producing flowers, if possible,

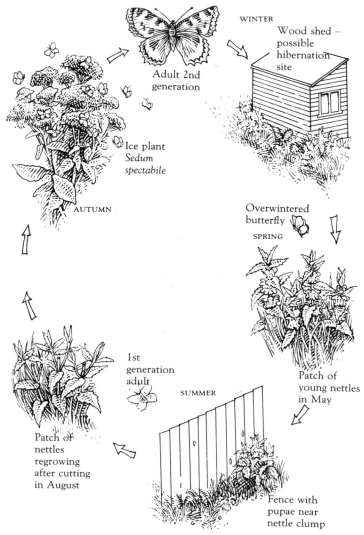

The drawing above shows the way in which this butterfly uses the different features of a garden during its life-cycle.

Small tortoiseshell butteflies lazily sunning themselves and feeding on Sedum spectabile (ice plant) nectar in the last few weeks before they hibernate for the winter. The caterpillars (left) feed on nettles in large groups until they mature.

A red admiral butterfly investigating a fallen Victoria plum. This species is particularly attracted to sweet fruits in autumn.

and, in due course, a nettle-patch in which to mate. After mating, the females disperse to seek out specific clumps of stinging nettles on which to lay their eggs in clusters of 100 or so. However, not just any nettles will do, and the butterflies tend to seek out plants with young growth in warm sheltered sunny places. After about a month spent feeding as caterpillars and now fully-grown they disperse to find suitable places to pupate (turn into a chrysalis). For this they need a range of fences, walls or reasonably sturdy vegetation. The pupae will turn into adults quite quickly, so that this generation will also mate and lay eggs in the same summer, and it is their offspring which hibernate through the following winter. But, by midsummer, there is much less young nettle growth for egg-laying, and clumps which have been mown or cut back and are regrowing will prove particularly attractive to the females if they are in a warm sunny place.

Once you start looking into the requirements of one insect, you soon realise that they are much wider than, for example, the presence of one plant species such as *Buddleia*, which may have drawn your attention to the butterfly in the first place. The more we discover about any insect, or any other species for that matter, the more it turns out that their requirements are complex and varied, often with different needs at each stage of

their life. Of course, with mobile insects like the small tortoiseshell, it is not necessary for them to find all that they need within a small area like one garden, but for many more sedentary species, the closer together their special requirements are, the better they will fare. There are plenty of examples from the wild of butterflies, such as the smaller fritillaries, dying out in an area when conditions have become unsuitable, even though there were suitable places only a mile or two away waiting to be colonised.

Amphibians, such as frogs or newts, provide another quite different example of the complex requirements of nature. It is not sufficient simply to provide a pond and assume that these species will appear. First, each species has different, and sometimes quite specific, requirements for conditions in the pond. They will also be greatly affected by factors such as the presence of fish in the pond, since the spawn and tadpoles are highly vulnerable to predators. In due course when the young froglets, or 'newtlets' emerge and disperse, if your pond is set amongst a large area of concrete or other open hostile land, most of them will fail to make it across this barren waste, as they will be picked off by waiting predators. They will also need hibernation sites, a constant supply of suitable food, places to hide during the day, and eventually a reasonably easy route back to the breeding pond for mating and eventually laying their eggs.

Whatever the species of animal, the different areas of your garden will interact to provide all or part of its needs. The better the relationship between these areas, the more likely you are to attract and retain a range of wildlife.

Species fit together

Not only does each animal that could occur in your garden need different aspects of it, but all the various plants and animals in the garden interact with each other and depend on each other.

Any biological system starts with the plants, for plants are the basis of all life on earth, as well as in an individual garden. They are the primary producers – the only things able to convert sunlight into living material. The whole animal kingdom, whether predatory centipedes, plant-feeding aphids or the omnivorous fox, depends ultimately on plants. If your garden was made up wholly of concrete, on which you placed some bird food and a vase of nectar-bearing flowers, you would probably attract a small range of common birds and a few insects, such as hoverflies. But they would have to rely entirely on other gardens or areas of countryside for the remainder of their needs – for nesting, mating, roosting and hibernating. If all the gardens in the area were solely concrete, there would be nothing to attract regular visitors. Long-distance travellers might occasionally drop in but these would be few and far between.

As soon as you put living plants into the garden, the position changes – thank goodness! The herbivores such as aphids, leaf-beetles, sawfly larvae, and slugs, and even a few mammals which feed directly on the growing plant material, will be able to survive. Other species, particularly insects, will feed as adults on the nectar or pollen produced by the flowers, gaining energy for flight, dispersal and mating, and possibly staying in the garden to lay their eggs, if suitable places exist. The fruits and seeds, though intended primarily as the way in which the plant disperses and reproduces, also provide high-energy food for many species. Goldfinches, for example, are well known for visiting thistle seed-heads or teasels, while blackbirds feed on raspberries, apples and any other fruit that they can find. A little lower down the scale some butterflies, such as red admirals, feed on rotting plums, while many larvae develop in the protein-rich environment of maturing seed-heads – for example, holly blue butterfly caterpillars feed on the developing fruits of holly and ivy.

Down below, there are many decomposers, which live on the plant material that has dropped to the ground and begun to rot. The rotting process itself is hastened by many fungus species, most of which are unseen; some however, such as mushrooms, put up visible fruiting bodies at certain times. Earthworms, millipedes and many other subterranean invertebrates feed wholly on the decaying detritus found living below the plants.

This great mass of animal life that feeds directly on plants, in one way or another, falls prey to a huge variety of predatory species. The predators are really just as dependent on the plants to produce populations of their prey as are these herbivores themselves. Plant-eating invertebrates tend to grow and reproduce quickly because their food is usually readily available, and does not have to be sought out. Aphids for example – including the much despised greenfly and blackfly – reproduce at a phenomenal rate. They have, in effect, no natural defences at all but rely for survival on their rapid reproduction rate. They do not even move away from an approaching voracious ladybird. Those that are left will carry on producing new young and, in good years, this production greatly exceeds the rate at which predators can eat them!

The opportunities for predators, even in somewhere as small as a garden, are endless and the number and variety that occur is surprisingly large. There are the large obvious species, like the song thrush which eats the snails, the spotted flycatcher which catches many different insects, the slug-loving hedgehog or the blue tit with a beakful of caterpillars gleaned from a decimated leaf. But there are many less obvious ones, too, such as the nocturnal ground beetles that prey on slugs and other invertebrates, or the *Dysdera* spider which emerges at night to catch woodlice. Or, the mass

The larvae of both ladybirds (left) and lacewings are voracious aphid-eaters, and they spend most of their time seeking out and consuming them.

All small garden birds feed their young on insects. This marsh tit has brought a beakful of moth caterpillars back to the nest.

of web-spinning spiders, like the garden spider, with their gossamer traps strategically placed across well used insect flight paths. Common wasps, though best known as pests in autumn, when they turn to seeking out sweet substances, spend most of the summer collecting living insect prey for their larvae.

Higher up the food chain still are the predators which eat other predators, but these tend to be fewest in number forming the tip of the numerical pyramid which has the mass of herbivorous species at its base. If you are lucky, you will have sparrowhawks, weasels and other 'top' predators in your garden, but they will not stay long, or come very often, just creaming off the highest layers of your garden's productivity.

Choosing the right plants

It is one thing to point out that plants form the basis of a wildlife garden, but it is another to decide which plants to use and where to put them. The later chapters give more details on exactly which species to choose for each situation, but there are some interesting general points to consider on how plants affect wildlife. If, for example, your garden was planted with a dense carpet of only one species of *Hosta*, it would be very dull for wildlife. Very few insect larvae eat *Hosta* leaves, the flowers are not particularly attractive to insects, there would be virtually no scope for nesting birds and cover for only a limited amount of ground-dwelling invertebrates. If, however, your garden was covered entirely with hazel, then the situation would begin to look a little better. A good many moths and other insects (though no butterflies) will eat hazel foliage or fruits in their larval stages, so the variety would begin to develop. A few wild plants could begin to colonise under the hazel, giving a little more variety still. The bushes themselves have rather more structure than the *Hosta*, so that several birds could find suitable nesting-sites, and even a few mammals could move in. The nut harvest, each autumn, would inevitably attract a further range of mammals, birds and invertebrates, and already we can see a more interesting picture emerging.

But why should hazel be better than *Hosta*, and what

THE BODY SNATCHERS!

The parasites, are a largely unseen and unknown group. Many of them are invertebrates, particularly insects, that live on other creatures while they are still alive. You may, for example, have noticed occasionally that a butterfly pupa will appear somewhere unusual, such as at the top of a fence post. A little later, instead of the butterfly, a mass of furry yellow oval 'cocoons' appear; these are the pupae of a little Braconid wasp, which have eaten the butterfly pupa from the inside. Another reasonably familiar parasitic insect is the yellow ophion Ophion luteus, a large gangling 'fly' (actually an ichneumon wasp) which is often attracted to lights at night. The females lay their eggs in the caterpillars of certain moths, which are eventually killed by the developing ichneumon larvae inside them. Interestingly, the adult ophions are one of the main food items of some bats, completing another link in this nocturnal food chain. These parasites are often abundant in the garden, their numbers rising and falling as the numbers of their host species change.

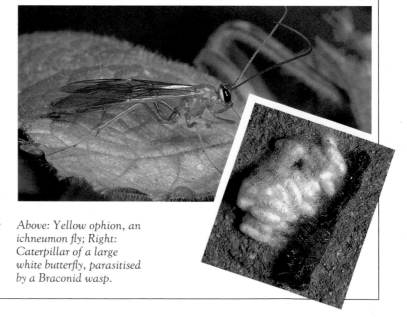

Above: Yellow ophion, an ichneumon fly; Right: Caterpillar of a large white butterfly, parasitised by a Braconid wasp.

A beautiful example of a garden managed on organic principles with wildlife in mind. The idea of this demonstration garden in Germany is to create a natural balance of predators and prey, in the complete absence of any chemicals. The combination of a series of miniature natural habitats attracts masses of wildlife.

makes one species more useful than another? There are several factors. Firstly, a native plant supports more herbivorous species of insects than an introduced species. Hazel scores over *Hosta* in this respect, since it is a long-standing resident of northern Europe and has built up its own varied invertebrate fauna. *Hosta* may have many insects feeding on it in its own native land, but these are not present in European gardens.

Secondly, the shape and structure of plants is important, which relates to the way in which you manage them as well as to the type of plant involved. Hazel is a woody shrub which is of sufficient size and form to be useful to a range of larger creatures, while clearly *Hosta* has no such advantages. A larger plant still, such as an old oak tree, offers even greater opportunities and has more colonisers as a result. Big,

---

GO NATIVE!

It is a good general rule to choose native plants for the garden, whenever you can, because of their ability to support a more varied insect fauna and generally, you need not be concerned that these insects will eat your other garden flowers for the very reasons outlined.

Some insects are very catholic in their choice of larval food-plant. The familiar cuckoo spit, for example, occurs on several hundred different plants. Most others are very specific to a group of plants, or even one species, and quite often require that species to be in a particular state before it is of use.

SHAPING UP

Woody plants can be managed in different ways to alter their shape and structure. A shrub cut down to ground level will produce a mass of new growth, giving rise to a broader, denser bush than the original single-stemmed version. Similarly, if you cut the branches of a shrub back towards the main stem, you will promote more new young growth than if you leave them untended.

---

mature trees have holes, dead wood, massive forks and fissured bark, as well as a much greater area of foliage.

Thirdly, the type of flowers and fruit that a plant produces will greatly affect the animals which are attracted to it. In this case the country of origin is irrelevant, as an exotic plant produces nectar which is just as available to insects as that from native flowers and, indeed, many of the best nectar-sources in gardens are not native plants. *Buddleia*, for example, is famous as a butterfly bush, but is native to China, and there are many other examples. Similarly, edible fruits of any sort, once discovered, are utilised by visiting animals.

It is therefore important which plants you choose for your garden, as well as how you manage them and where you put them, and the choice of species given later tends to reflect this. Having looked at what is going on in the garden, we can now turn to looking at some of the features you need to consider when designing your own wildlife garden.

WILDLIFE
BY
DESIGN

For many people, the term wildlife gardening conjures up images of a weedy wilderness. You may have spent a lifetime trying to get rid of the nettles from the bottom of the garden and the weeds from the lawn, only to be told now that you should have kept them all along. Some find it hard to reconcile wildlife and their idea of what a garden should be. In fact, the joining of the two ideas together in the term 'wildlife gardener', gives us a clue to the way in which this type of garden can work in a variety of different settings. First and foremost, the wildlife gardener is just that – a gardener, and although they may not describe themselves as such, all gardeners are in fact environmental managers. Although a garden may seem more or less natural to us, the person tending it will know that it is really a carefully controlled environment. The tightness of the control may vary, but it is always there. One has only to think of the time and effort that goes into so many lawns in order to maintain a pure stand of grass at an even height; the intervention of a two-week holiday can show us the effects of failing to exercise one management technique for just a short time. The efforts of the wildlife gardener are directed towards achieving a much richer environment for plants and animals, but within a garden context.

The potential for wildlife in gardens like these opposite, is tremendous, even in the centre of a town.

Planning your wildlife garden

At this point, it might be helpful to consider what it is that makes a garden, even a wildlife garden, different from a nature reserve. Of course, there may well be some that are combinations of both, but in general there are several important differences. Most nature reserves are fairly sizeable, usually rural in location and have been established because of some existing natural interest. Although reserves increasingly have to cope with other pressures, their primary function is normally one of conservation. The garden, on the other hand, *may* be large and rural with a colony of rare bats in the cave below the meadow, but it is just as likely to be extremely small, in the middle of a town, and devoid of much of anything at all. Add to this the fact that gardens may be expected to double up as children's playgrounds, outdoor dining rooms, drying yards, storage areas for all manner of things and suppliers of produce (to name but a few) and still look good, we can see that the differences may well be considerable. When we make our garden into a garden for wildlife, we have to arrange it so that these other functions can continue, and we hope that the garden will also be a beautiful and pleasing place to be in.

THE DESIGNER WILDLIFE GARDEN

One can consider the design of the wildlife garden in very much the same way as that of a more conventional one. The end result is likely to be effective from all points of view, including that of the wildlife, than that produced by a more hit and miss approach. We said in the first chapter that diversity, both of habitat and species, is of paramount importance. In wildlife gardens, we often increase diversity by bringing together several small habitats which would not normally all occur together. It is easy to imagine that the results of combining many different elements in one smallish area might not always be as pleasing visually as one would have wished. But start with a positive design that you can incorporate these habitat elements into, and the end result will hold together in a very satisfying way. Designing a garden is a process which can be undertaken at all sorts of levels, according to inclination, and it does not matter whether you are embarking on an entirely new garden or making a small alteration to an existing one. Even at its simplest, the process encourages the designer-gardener to think about what he or she is aiming for and how best it can be achieved. You may not need or want to follow all the suggestions in the rest of this chapter but, by just reading and thinking about those ideas which apply to your garden, the end result must benefit from the thought that you have put in.

Above: Listing your aims and intentions will help you cater for the everyday activities, even in a wildlife garden.
Above left: Taking stock of a garden's potential – here, shadows show the aspect of this long thin plot; flats dominate; and other gardens crowd around; but hedges and trees have good wildlife possibilities.
Left: The type of habitats you create in your garden may be influenced by factors other than ecological ones. The 'wildlife' shown here needs to be attracted away from some areas!

The starting point

Firstly, you need to think about your starting point, whatever it may be, an old established garden, a long neglected wilderness or a bare plot on a new estate. A list of your garden's good and bad factors is the first priority, as this may well affect your eventual aims. This, and other parts of the design process can be worked out mentally, but, for most people, it does help to put pencil to paper. A more visual way of recording what you have involves making a plan of the garden which you can use to base your ideas upon. Whether you do this or not is very largely a matter of individual preference; some people are happier planning out a garden on the ground, while others find it easier to work things out on paper. There is no need to be an artist, as even a rough sketch can be very helpful. However, an accurate plan is not difficult to make. It can prove invaluable for working out ideas, and deciding the relationship between the different aspects of the garden. It also helps you to look at your garden in a new way, without the encumbrance of what is already there. It is possible to make a copy of the large-scale

Photographs are very useful, both as a reminder of what your garden is like, and as an aid in working out what you want to do. Laying tracing-paper over a snapshot, and making a quick sketch of your ideas, can help you to imagine the final result.

local map of your area at your local planning offices, as long as this is for your own use. Instructions to help you make an accurate garden plan are given on page 36.

Facing the right way

At this early stage you will also need to record such things as the aspect of your garden (the direction in which it faces), because the positioning of some features will be dependent on this. Not only will you want to ensure that your own sitting area is not in total shade for most of the day, but you may wish to site a grassy bank facing the sun or a rotting woodpile in the shade. You will need to be aware of places which might be especially favourable or difficult for plants – a warm wall or a frost pocket – and you need to bear in mind the general climatic area in which you live.

Different soil layers show as different colours in this typical heathland soil profile. Digging a hole in your garden could be equally revealing!

There are a few thoughts to bear in mind if your garden is afflicted with either canines or juvenile humans. One, of course, concerns ponds and young children, as it is difficult to make a pool completely safe, yet still attractive. The reverse problem, of potential damage inflictable on a pond, is not often foreseen. The dogs which cannot resist jumping in, or the ecologically-aware children investigating a dragonfly nymph with a handy stick, may both cause a puncture in a pool liner. If you are planning a meadow area, dogs (the jumping on imaginary things in the grass variety) and children (the sort which entail bikes and football and dens in the shrubbery) must again be considered. Unless you can provide adequate space for both, bitter experience indicates that your cherished meadow will be permanently flat.

Down on the ground

You will need to know the soil type (whether it is acid or alkaline – the pH – and whether it consists of sand, loam, silt or clay – the texture). A very acid or alkaline soil restricts the range of plants that can be grown and, although it is possible to change this to a certain extent, this can involve a lot of expense and effort; it is usually best to plant according to the soil that you already have.

You can test the pH quite reasonably and accurately with a small kit (not the probe type). Even in a small garden, you will need to test the soil in several different areas as it can vary considerably. Do not use the top layer for these tests – the results will be misleading – unless your soil is very shallow, you need the next spade's-depth down. If you can dig a even bigger hole, you will find out even more about the potential growing conditions in your garden. A sandy soil might have a hard layer of iron pan that you would have to break up before planting any large trees and shrubs; a clay soil might show the tell-tale blue-brown mottling which comes from frequent water-logging. You will also be able to see if water rises or stands in the hole, and how quickly it drains away. You are probably well aware of the texture of your soil if you have worked the garden for any length of time but, if not, feeling the soil in your hand will give you a rough idea. The grittier it is, the more sand it contains; the stickier it is, the more clay. You can also shake some up with water in a jam-jar; the biggest, gravelly particles sink to the bottom, then sand, then clay, then humus – decomposed organic matter – on the soil surface or floating.

This beautiful 'nectar' border is intended to attract insects, but its visual appeal owes much to the care which has gone into its design.

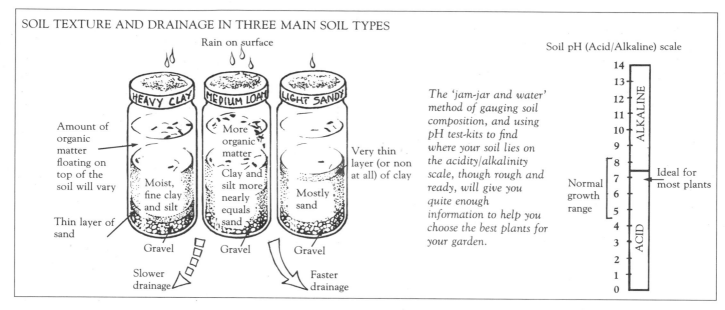

SOIL TEXTURE AND DRAINAGE IN THREE MAIN SOIL TYPES

Rain on surface

HEAVY CLAY   MEDIUM LOAM   LIGHT SANDY

Amount of organic matter floating on top of the soil will vary

More organic matter

Clay and silt more nearly equals sand

Very thin layer (or non at all) of clay

Moist, fine clay and silt

Mostly sand

Thin layer of sand

Gravel   Gravel   Gravel

Slower drainage

Faster drainage

The 'jam-jar and water' method of gauging soil composition, and using pH test-kits to find where your soil lies on the acidity/alkalinity scale, though rough and ready, will give you quite enough information to help you choose the best plants for your garden.

Soil pH (Acid/Alkaline) scale

14
13
12 ALKALINE
11
10
9
8   Ideal for
7   Normal   most plants
6   growth
5   range
4
3   ACID
2
1
0

The soil type affects such things as drainage and the ability to hold nutrients, and thus the plants that will do best. It is the clay element in the soil that holds the nutrients available to plants, and thus a clay soil is usually a fertile one, although one with a great deal of clay will not drain very well and may crack in summer. Loams have a more or less equal balance of clay and sand, giving a reasonable level of fertility, and will normally have a fair amount of humus. Sandy soils normally have little humus and, although their low fertility will suit a wild meadow, you will need to add organic matter (compost or well rotted manure for example) to parts of the wildlife garden, just as you would in a more conventional one. Apart from the information in this book, you can check on the suitability of plants for different soils when you buy, and also look around you to see what grows well in the gardens and countryside surrounding you.

The world outside

The last and maybe the most important fact that you need to consider is the relationship of your particular garden habitat to the wider one around. Good groundwork at this point should help to ensure that your garden can make the best possible contribution to the surrounding environment, and if you are really enthusiastic you could make your own observations on the local wildlife. You may well discover a new hobby! You will also have a better idea of what you might reasonably expect to attract into your garden. One suggestion that has been made is to make an environmental map of your locality. You will need to reverse the importance normally given to such structures as

buildings and emphasise areas of known or potential wildlife value. Doing this will help you to see how your garden fits into the wider network of areas available for wildlife. Even if your garden appears to be totally isolated, you may be spurred on to provide a wildlife oasis in an urban desert, but you will usually find that even the most urban areas have a fair sprinkling of green. Other sources of information are the local natural history or conservation societies – your local Nature Conservation Trust or the local group of the Royal Society for the Protection of Birds are examples.

The wildlife value of a large garden, which can accommodate several complementary habitats, may be much greater than their individual values added together.

Stating your intentions

By now, you will have done a lot of work finding out what you already have. The next stage means another list – one of intention. This is going to be a statement of what you (and your family, if you have one) hope to achieve within the boundaries around your home. It will depend to some extent on your motives; conservation, an interest in learning about nature, your own enjoyment. This is not as daunting as it may sound; all you have to do is list, usually in the form of actual things that you want in your garden, the functions you would like it to perform. It does not matter whether you are changing part of an existing garden, when you will probably only consider a few things to do with that small area; or if you are starting from scratch, in which case the list will probably seem endless – this is a list of possibilities! It may include such requirements as space

for the barbecue or the washing line, and preferences for individual plants or features. Since this is to be a garden for wildlife, this is when you must think about what your aims are in that respect and what you are going to put into your garden in order to achieve them. Your motives may be shaped by an interest in one particular field – ornithology perhaps – and this will determine the type of wildlife features that you will want to put into your garden.

These lists have the added advantage of being easy to compare, and also to check to make sure that your proposals do not clash with each other, or include any wild impossibilities.

A pond or a meadow?

A small change to an existing garden will probably only involve the introduction of one or two new habitats; this should be quite straightforward. However, if you are contemplating major changes, perhaps the main choices that you will now be considering will be those concerning the habitats you might incorporate. This

Environmental maps are surprisingly revealing. Compare the total green-space with areas of really good wildlife potential and potential garden wildlife space. Look at 'green corridors' in, and leading into, the area.

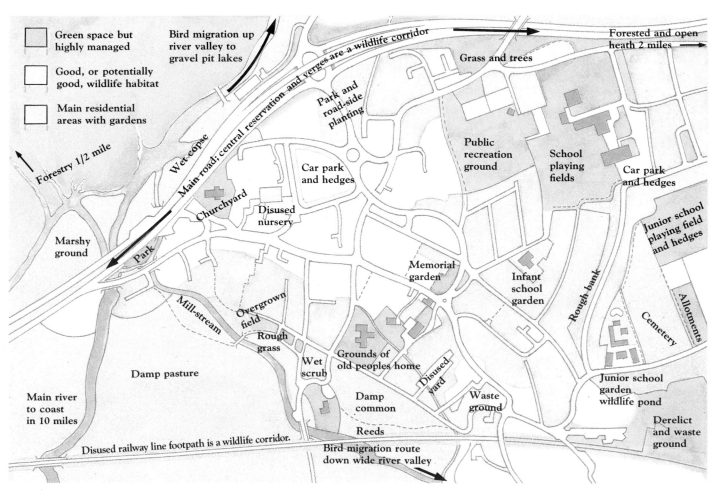

will depend on the nature of your garden, its size and its surroundings, as well as your personal preferences. A large garden can accomodate fairly big areas of different habitats, but these could be very much reduced in the small garden. This type of mixed-habitat small garden is certainly valuable for many species and is lovely to look at, but the likelihood of success with a wider range of wildlife will not be as great as in the larger garden.

Different habitats vary in the size that they need to be for maximum ecological usefulness. A pond, especially if associated with an adjoining marshy area, can be quite small and still be a surprisingly wide range of wildlife. However, although even a tiny patch of flowery meadow can look beautiful, and this alone may make you want to grow one; if you are hoping that some of the grass-feeding butterflies will establish in your garden, you will need to provide a reasonable area of grass in a mosaic of different heights – long, medium and short. Although this need not be huge (minimum $2\frac{1}{2}$ by $2\frac{1}{2}$ metres – 8 by 8 feet) the chances of success do increase with size.

For most of the habitats mentioned in this book, the relationship between size and wildlife value is fairly obvious, although it should be borne in mind that it is often the transitional areas between one habitat and another that are especially valuable. In many gardens it would be quite reasonable to concentrate on the creation of one main habitat. The value of this area is then enhanced by one or more complementary habitats, which do not need to be as large. In practice, a garden might consist mainly of a large meadow grassland area, but it could also be bounded by a thick mixed hedgerow; it could incorporate a small area of native shrub planting and maybe a pond with a small boggy area. Such a garden would fulfil the aims of the good wildlife gardener to provide both open areas and shelter, and the pond – as well as contributing its own value as a habitat – wildlife value acts as a booster to the other habitats. Whether the choice is for several habitats or for a single dominant one, one comes back to the importance of diversity in the wildlife garden. The mixed-habitat garden has an automatic high level of diversity within it, not found in a garden based primarily on one habitat type, although the provision of complementary habitats will greatly help. In a garden which is dominated by one type of habitat, you have to build in diversity by including different elements within that. For instance, a woodland garden might have an area of dense shelter planting opening onto a more grassy area, log piles in different situations, a damp hollow and a shady bank. You will have to bear in mind the needs of these different habitats and the

DESIGN CONSIDERATIONS FOR · THE WILDLIFE GARDEN ·

Requirements
● essential/advantageous
○ possible

	sunny or unshaded	shaded	sheltered	level	low fertility	near water supply	near house	in wild area	near other habitat	time needed
Woodland edge	○	○	○		○			●	●	●
Hedgerow	○	○			○		○	○	●	●
Log pile		○	○				○	○	●	
Compost heap		●	●	●			○	●		
Bird-box		●	●				○	○	●	
Bird-table	○	○	●			○	●	○		
Flower lawn	●	○		●			○	○	○	
Wildflower meadow	●	○	●		●		○	●	●	●
Butterfly meadow	●		●					●	●	●
Nectar border	●		●		○	○	●		○	
Cornfield patch	●		○		○	○	○	○		
Herb bank	●		●				●		○	
Rocky outcrops	●	○	○		○		○	●	●	
Banks and walls	○	○	○				○	○	○	
Heathland lawns	○	○			●		○	○		●
Rocks/rubble heaps	●	○						●	●	
Pond	●		○	●		●	○	○	●	
Raised pond	●		○	●		●	●		○	
Wetland	●	○	○	●	○	●	○	●	●	

Adapted with kind permission of BTCV, London

elements within them, when you decide what to include in your wildlife garden. If this is new to you, the chart above will remind you of different possibilities and their requirements.

Taking life easy

Wildlife gardens are sometimes recommended as a solution to the ever-present problem of upkeep. This can certainly be one of the additional benefits of such a garden, but it is by no means an automatic result. The amount of work entailed will depend very much on the type of habitats that you are thinking of creating in your garden, as low maintenance techniques will apply more

easily to some than others. If you think about the easiest gardens to look after, these components might be: hard surfaces, trees, shrubs with ground cover and/ or mulches, and grass (though it takes time to do the mowing, it is not a difficult task). These do have their equivalents in the wildlife garden: woodland edge, native shrubs, flower meadow – low maintenance once established – but if you decide that you want a wildflower butterfly border you may well find that it is as much work as a traditional herbaceous border.

Suiting both the habitats and the plants within them to your garden is important not only for the wildlife. If the backbone planting of the garden is native to your local area, you have the great advantage of knowing that it should do well and not need a great deal of cossetting. This is part of a low maintenance approach that does apply to the wildlife garden, letting nature take its own course – the gardener guiding from behind rather than pulling from the front. On the ground, this might mean giving shrubs rather more growing room when planting to allow them to reach their natural size without a great deal of pruning, or letting plants seed freely around the garden and removing those which are not wanted before they grow too big. Both of these examples require a degree of compromise; the unpruned shrubs may carry somewhat fewer flowers – letting plants self-

'Our green and pleasant land' can be surprisingly sterile from a real bird's eye view, but we can change the picture by designing wildlife-greenspace into our gardens.

seed means coping with extra weeding. In this freer form of gardening, it helps to know the plants and their potential, and, if you are going to let wild flowers self-seed around your garden, you will have to learn to recognise their seedlings – a guide to weed seedlings is a good starting point. There may seem a lot to learn at first but, like most things, it gets easier with practice.

Ways and Means

One last very simple way to almost magically reduce hard work in the garden is to think about whether you really do need to do the job in question, either in that way at that time or, perhaps, at all. If the answer is no, it may well be that there is an easier alternative. For example endless weeding can be saved by covering the ground with anything you can lay your hands on. Another option for the dedicated low maintenance gardener is to resist the urge to tidy up whenever possible. Keep tidiness for a few areas, maybe the flower-beds near the house, that really matter and that you enjoy most. This is an approach that suits the wildlife garden down to the ground.

Making a garden base plan

You can make a garden plan as simple or as detailed as you like, as its function is to help you decide how to arrange your garden. You may find that your garden is somewhat different from what you had always imagined. Drawing up a plan is not a difficult procedure.

You will need the following:
Large sheets of paper – a pad of layout paper is ideal;
A long tape measure – you can pace out the distances, of course, but it is more difficult to get to awkward areas in this way;
Marker pegs for measuring to in the garden – tent pegs work well;
Another person – measuring is much easier with a helper.

Unless your garden is very simple, you will probably get in less of a muddle if you use a new sheet of paper for each set of measurements.

Start by drawing a large outline of your house, or the part of the house adjoining the garden, on the first sheet of paper, and mark in the doors and windows. Now measure round the house, stopping at each corner that you get to. You can record the exact positions of doors, windows, patio etc. if you wish. Note by arrows any particular views from the house or its surroundings that you wish to either emphasise or hide.

Now take a clean sheet of paper, and draw a sketch plan of the whole garden with the house in position. This drawing does not have to be a masterpiece – if you can understand it afterwards, that is all that matters. Use the tent pegs to mark the points that you are measuring to; this is much easier than remembering which side of a metre wide hedge you went to with the last measurement.

You need to measure as follows:
From the corners of the house, measure to the different corners of the garden. Aim to fix each corner of the garden from at least two different points on the house. If your tape is not long enough to reach the whole distance use marker pegs to divide the garden into manageable sections, and measure those. You may, for example, find it easier to divide a long narrow garden into several squarer parts. If your garden is reasonably simple, this may be all you need to do to fix the basic outline of the garden. You can increase the accuracy of the final plan by taking some further measurements.

Again from the house corners, in a straight line with the house walls, measure the distance to the garden boundaries. Measure all round the perimeter of the garden, using the marker pegs for reference.

You need another sheet of paper

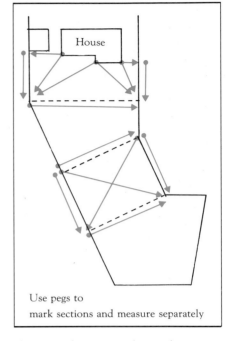

Use pegs to mark sections and measure separately

if you wish to note down the positions of any other features within the garden. Include the positions and type of trees, shrubs (in a group will do) and hedges. If you want to do this roughly by eye, it is easier to wait until you have an accurate plan of the garden. You can make exact measurements either from two fixed points, as you did for the corners of the garden, or by measuring the distance at right angles to a point along a known line and the distance of that point from one end. On this plan, you should also include the north point, to remind you of

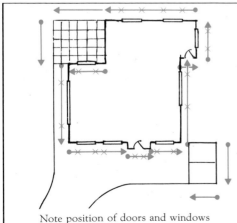

Note position of doors and windows

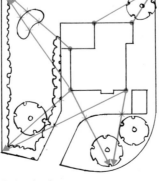

Record other measurements as you go along

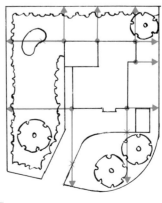

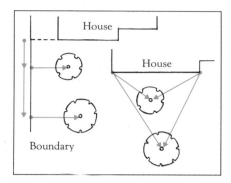

the general aspect of the garden, and any other noteworthy features of the garden, for example areas of permanent shade or bad drainage. You may also need to mark in positions of drains or soakaways.

Mark in any major changes of level in the garden, noting which ground is high or low, and estimate by how much. Absolute accuracy is not necessary unless you are planning a lot of constructional work. If you want to work out a particular area in more detail, draw that area again on yet another piece of paper.

Drawing up an accurate plan

It is simpler to draw an accurate plan on a squared base, so either draw directly onto a large sheet of graph paper or use a sheet of tracing paper over the top. The latter method enables you to correct mistakes more easily, and also to remove the squared paper when you reach the design stage, and might find it inhibiting.

The aim now is to produce a plan of your garden that is to scale. You will be able to use it to help work out the changes that you want to make and to place new features by working back from the plan to the garden. Whether you have measured your garden in metres or feet and inches, you should make the final plan to an appropriate scale. For example, a scale of 1:100 means that every 1 centimetre on the plan represents 100 centimetres (1 metre) on the ground. A scale of $\frac{1}{8}''$:1′ means that

each $\frac{1}{8}$ inch on the plan represents 1 foot on the ground (or that each inch represents 8 feet). The scale you choose will depend on the size of your garden and on the size of the paper you use.

A2 paper (59.4cm × 42.0cm or approximately 24″ × 17″) is a good size to choose, as it is big enough to show an average garden at a reasonable scale. A1 paper is twice the size of A2, A3 half the size.

On A2 paper

At scale 1:50 (approx. $\frac{1}{4}''$:1′) you can fit on a garden of about 29 metres × 21 metres

At scale 1:100 (approx. $\frac{1}{8}''$:1′) you can fit on a garden of about 59 metres × 42 metres

At scale 1:200 (approx. $\frac{1}{16}''$:1′) you can fit on a garden of about 119 metres × 84 metres

You will fit in slightly more yards if you have chosen to measure in these.

Keeping in scale, start by positioning the house on the paper, referring to the rough measurements to make sure that you have enough room to fit the garden round it. You may find that when you draw in the house, it is impossible to make all the measurements fit. Unless a distance is wildly out, this is probably because you have assumed the corners of the house to be exact right-angles, and even modern houses very rarely are. In this case, you will need to "adjust" the least important measurement to fit, perhaps the side of the house along an access path.

Now draw in the garden. Measure out from the house on the plan in the same way as you did when measuring on the ground. If you have compasses, then it will be easier to fix the positions of the corners of the garden that you measured from two corners of the house. Check what you have drawn against the other measurements that you have taken. At this stage,

it may be helpful to go back and check a doubtful measurement in the garden, or take a new one, but do not worry about slight discrepancies. When you have worked out the changes you want to make in your garden on the plan, you will then transfer any measurements for these from the plan to the ground, using the house as the reference point in the same way as when you transferred measurements from the garden to the plan. This means that any slight inaccuracy can be accommodated at the edges of the garden where it will matter least.

Finally, transfer the rest of the information from the rough drawing to the scale plan.

You should now have an accurately scaled plan which you can use as a base for working out any new designs or changes in your garden. Do not draw on the plan itself, as it would be difficult to try out new ideas without spoiling it. If you work on tracing paper laid over this base plan, it can be replaced when it becomes too messy.

If you wish to enlarge all or part of your plan, you can take it to be photocopied on a machine with an enlargement facility (this will be from A4 to A3), and ask for it to be enlarged twice. This will double its size (halving the scale) with sufficient accuracy for most purposes. Do bear in mind that pencil lines may not show up well enough to be copied, and may need to be drawn over with a black spirit pen (blue lines will often not copy at all). If the plan to be photocopied is on tracing paper use a white backing sheet to show up the lines. The plan will also need to be copied in sections to fit on the machine, and the net result will be several pieces of paper which need to be taped together. Because of the copying process, the lines on adjoining sheets may not tally exactly, but the end result will be good enough to be useable.

A natural pattern

Y ou should now have a considerable amount of information, either in your head or on various bits of paper. What should you do with it? The end result will be a garden which achieves your aims, while making maximum use of what you had originally. You hope that it will be a sanctuary for wildlife, a garden that works practically, and that it will be aesthetically pleasing.

The next section gives you guidelines for working out a basic design, while considering some special aspects of the wildlife garden. For simplicity's sake, the suggestions are based on the use of a plan to design the whole garden, but you should still find them helpful if you are working on only a part of it, or without pencil and paper.

The starting point for this stage will be either the garden itself or the plan that you have made. The lists of what you already have and what you hope to achieve will be used as input for the design process. It is best to work on a sheet of tracing paper taped over the base-plan. When this becomes congested with discarded ideas, it can easily be replaced with a fresh sheet. You may have drawn your base-plan on graph paper anyway but, if it is on tracing paper, you may like to lay this over a grid-sheet (such as graph paper) to help you

The flowing lines and the balance between the open and planted areas gives this garden harmony which is pleasing to the eye.

structure you design. You will also need a soft pencil (at least a B) and a soft eraser. There is no need to use a ruler as precision draughtsmanship is not called for. Your design is likely to flow much better without one, but there are many people who do find it reassuring to use a ruler. At this point, it is worth repeating that drawing ability is not at all necessary – enthusiasm for what you want to do is far more important.

Pencil to paper

Put your working sheet of tracing paper over the base-plan and trace off the outline of the house, garden, and anything else that will affect the design. This at least

The subtle spaces in this garden are defined as much by their vertical structure as by their ground-plan – an important point in your own design.

gives you some lines on the paper and stops it looking quite so intimidatingly blank. Put in the compass-points, N, S, W and E, and make them noticeable; this will remind you to think about sun and shade, and other effects of aspect. It is also useful to note any good or bad views.

Now you need to start drawing the actual plan. There is no obligation at this stage to produce the final design, or even get anywhere near it, and it can be as messy as you like (it probably will be). Once there is something on the paper it can be developed or changed in various ways, and you can experiment until you have done something you are happy with. Whichever way it is

arrived at, the success of a design depends very much upon the balance that it achieves. This is especially true for the wildlife garden because of the needs of the plants and animals themselves. Think of the type and position of habitats you wish to bring into your garden, and also their size in relation to each other.

While working out a design, consider the relationship of the garden to the house and the surroundings. Think about such aspects as what your house is built of and the materials in use locally for walls, fencing or hedges. Wildlife gardens need a natural basis so it is important to use harmonious materials. This does not mean that such a garden *has* to be informal. It is perfectly possible to have a formal design in a small town garden, for example. The formality will be balanced by the 'naturalness' of the planting and other ingredients of the garden. However, in suburbia or a rural location, an informal design is likely to be more appropriate, even if you can keep a certain formality near the house. Wherever you live, the chances are that the boundaries of your garden are straight, as they are artificial divisions superimposed on the landscape. A town design may honestly acknowledge this but you will probably want to disguise the edges of your plot in some way which is more in sympathy with the natural effect. The English Landscape gardeners of the 18th century used the ha-ha to give the impression of total continuity of the garden into the landscape, but a small modern garden is more likely to rely on the use of planting to soften the boundary line. It often helps to concentrate planting, especially of evergreen and taller items, around the corners. This will work even in a small plot and climbers can be used in between. A bad view can be hidden and a good view emphasised by the boundary planting on either side. This boundary

> ## A REFUGE FROM THE WORLD
>
> *The experience of a garden, from whatever standpoint, is only partially limited by its boundaries. This is even more true of the relationship of the wildlife garden to its surroundings, which form a small part of a much wider and more complicated system. The first gardens were sanctuaries defended against the outside world, and this idea persisted until times became stable enough to allow the extension of the garden into an idealised countryside. It is a strange reflection on man's activities that wildlife should now need sanctuary in gardens and that, in providing it, we have brought this idealised countryside into our suburban environment.*

planting will be part of the shelter provision for wildlife.

A *spatial relationship*

If you now turn your attention to spaces within the garden, it will help you think about the basic structure if you mark – just with a line – any fixed access routes

---

Two stages in spatial development of a design – final design outlines emerge from the simple shapes and lines which define areas and access on the first sketch-plan.

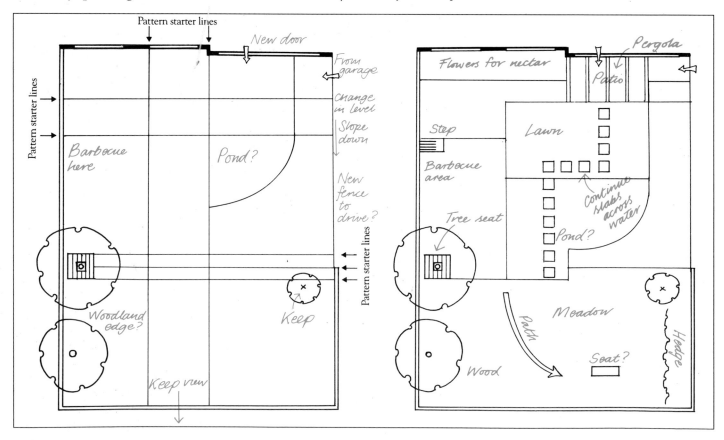

The relatively large area of water in this Chelsea Show garden (for the British Trust for Conservation Volunteers) was designed to impart an air of calm and tranquility to a small town wildlife garden. Despite its straight lines, the garden contains all the elements of a balanced wildlife garden.

(how you get into and out of and through the garden); other paths are often best left to develop naturally – they will soon appear where you have walked. Start thinking about the different areas of the garden and use simple outlines to show those parts of the garden that you feel 'belong' to the house and its immediate surroundings. Most houses need a patio or similar area and this extension of the living area from inside to out, especially if sympathetic to both house and garden, automatically helps to tie the house and garden together. Now add more shapes (circles will do) to denote other areas within the design. These may represent other functions than wildlife, and it is at this point that you can decide where things will be in relation to one another. One example of this is the need to site areas which may encourage slugs and snails away from areas where you would rather not have them. You need to provide areas where they would rather be, in the long grass or the hedge bottom as opposed to the vegetable garden, where you follow the standard practice of keeping it as tidy as you can. Some people find it useful to think of creating clearings for various purposes out of an imaginary woodland. They will be bigger or smaller according to their purpose, but at the moment they need only be labelled as perhaps 'meadow' or 'pond with marsh'.

REMEMBER THE FRONT GARDEN

If you have a front garden, you will need to decide whether it is to be part of your plan as well. Front gardens have less time spent in them, but they are the face that we show to the world and consequently differ from our more private domains. If you decide that your front garden is to be as valuable for wildlife as the rest of the garden, then it may be worth modifying it just a little so that, for example, you might include a flower lawn but not a full-blown meadow. It will partly depend on factors such as the amount of room you have, but you will want to pay extra attention to the finishing touches such as a mown edge to a longer grass area.

You will probably find it best to arrange the garden so that most of the functions which involve the family are near to the house, and the provision for the wilder inhabitants, small mammals perhaps, further away. There will be exceptions, however; a bird feeding area gives added pleasure where it can be seen from a window, and does not need a six-mile hike in the snow to fill up the feeders. A colourful and fragrant butterfly border will be enjoyed nearer the house; a quiet place to sit and watch and listen to small creatures needs to be away from disturbance. Such a 'looking and listening' seat may be where the hidden paths come to, right in

This picture of Winllan Wildlife Garden, taken by its owner, shows how revealing it can be to view your garden from above. Straight lines near the house grade into informal 'wilder' areas further away, while strong shapes give the garden its basic form. The top right is a wildflower meadow set away from the house away from the more ordered parts of the garden near the house. The pond at the centre of the garden is a feature which is approached through a gap in the beds running across the lawned area. The tree at the centre provides shade in the summer from which to view the goings on of the pond life. The planning of this garden has been thought out and developed over a number of years producing a garden that is kempt and practical as well as providing all the right ingredients to attract as diverse a population of wildlife as possible.

the heart of the garden, and if you think you might use it for nocturnal observations, you will want it to be somewhere sheltered and comfortable. You may decide to include a little bit of human habitat in your wild garden, and make a summer house in a secluded corner to double as a bird-hide. There is no need to specify individual plants in any areas. It is better to concentrate on broader definitions such as 'nectar plants', or 'dense shrub cover'.

A developing pattern

You may prefer to base your ideas on the development of a pattern or patterns within the garden. This approach often suits 'back-of-envelope' doodlers, especially as the actual features need not be represented at first. They can be allowed to grow out of the initial pattern. This pattern can be made to tie in with the salient features of the house and indeed it often works to develop it outwards from the house becoming increasingly more natural as you move further away. The other approach which could be tried is that of the kitchen-planner, using scaled cut-outs of various

features and moving them around on the plan until a satisfactory end result has been achieved. You will have to guard against producing a 'bitty' design with this method. This is also the time to think about the ways in which you can use the design to disguise or emphasise particular features of the garden. These effects usually work by encouraging the eye to see what is intended rather than what may actually be there.

Whether you thought of your initial layout as a pattern or not, try to see it in that way now. Most gardens can be thought of as consisting of two elements – the underlying framework and the material which clothes it. Just as bone structure is the primary determinant of our own appearance, so the 'skeleton' of a garden is the basis on which its final appearance is built. The framework of your garden is represented by your initial layout, but it needs to be strong enough to carry whatever you clothe it with. This is more, not less, important in gardens where nature plays a major part. Many of nature's own boundaries are gradual transitions from one habitat to another and are not normally sharply defined over a short distance. These 'edge' zones are rich ones for wildlife and areas such as woodland edge are very valuable in the wildlife garden but, when used in a garden setting where one of the aims must be aesthetic, a well-defined base will contribute to unity of design. At this stage, the design of the garden should show purpose in simplicity and a strong line (no wiggles). A vague framework will disappear in the final garden and over-fussiness will obscure the original intention. If the basic design works, the extra details can be added later. Looking at the design upside-down will often show up any problems and help you to decide if it is what you really like. You will need to check against your original lists to confirm that your design fulfils all that you want it to.

The next step is to imagine the whole design in 3D. Many people find this quite difficult but photographs can help you visualise the final effect. Put a piece of tracing paper over a photo of the existing garden and work out a rough sketch of the proposals from this. If

DOES IT WORK?

You will need to know whether the design works from a purely practical point of view. Can you get a mower from A to B? Have you put the washing line where the clean washing will be constantly decorated by the birds as they take off from the berried shrubs? You now need to see how your proposals work on the ground and one of the best ways of doing this is to lay them out in the garden. If you have been working without a drawn plan, you will have done this at an earlier stage. Mark out your proposals with canes and string, or a garden hosepipe to give some form to your ideas. Looking at the results from an upstairs window will help you to see the design as a whole, and it also helps if you can live with the design for a while before you rush out and implement it.

you have changes of level in the garden this is when you should work them out.

Adding the detail

When the overall plan has been worked out in this way, attention can be turned to details. Now you are sure that your various objectives will be satisfied by the general proposals that you have produced, you need to

Bold use of colour is demonstrated by this mass of ragged-robin in a garden by John Chambers at the Stoke National Garden Festival.

Two stages in developing a design from pattern – the initial framework derives from a rectangular pattern (though it could be curving and/or informal) based on site features.

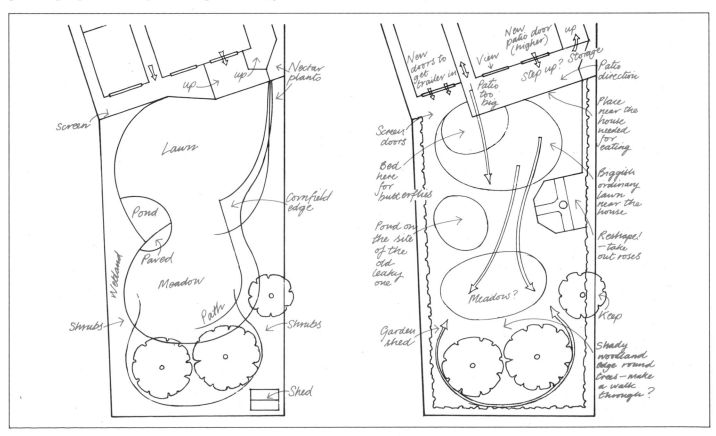

MAKE THE DIFFERENCE

Good general rules here are to try to avoid putting similar elements, especially in height, next to each other, and to maintain clearer boundaries nearer the house, for example between a well-mown lawn and longer grass.

repeat the process at a more detailed level. This will be easier because you already have a framework within which to work. It is only now that decisions have to be made on how to finish the edge of a pond, or the exact shape of the butterfly border. You should still be aiming for simplicity of shape and outline; an example from the conventional garden is that of the lawn – the permanent area of clear green makes an immediate, if sometimes subconscious, impact on the eye. The planting and other elements of the garden are set off beautifully by a simple shape of one colour, but all too often this area is nibbled away around the edges by flower beds and ponds until it looks more like a well-

Self-sown plants almost totally disguise the steps on the left. On a larger scale, planting will soften, but not obscure, the underlying framework.

Mown paths like the one below have a softness which is appropriate to the wildlife garden, and their position can easily be changed if required.

Butterfly borders owe much to the traditional cottage garden like this one in Oxfordshire, and its scents and colours are appreciated anywhere.

worn piece of jigsaw puzzle. Such considerations are no less important in the wildlife garden, especially when you are thinking about the boundary between one area and another. Also bear in mind the nature of these boundaries. Some may be best as gradual transitions both visually and for wildlife – the rough grass at the foot of the hedgerow or the wetland planting into the pond – while a clean division may be better for others – for the wildlife itself, for visual reasons, or for ease of management. You will probably appreciate a butterfly border more if it does not run straight into the wildflower meadow, for instance.

You will have already considered access within the garden, but now you need to think in more detail about access to various parts of it. Most of the wildlife you attract will be happier if undisturbed, but you, and others, will gain enormous extra pleasure from being able to see it. This simply means making access easy to some areas so that you do not have to crash through a patch of great willowherb to show a nephew a whirligig beetle. A path for this purpose need not be wide or hard-surfaced, or even permanent. It might be a mown path through the meadow, or wood chippings under

A Place for Wildlife in a Town Garden

This is the smallest garden that we have featured, and in the design sense at least, the most different from the others. The idea is to accommodate all the features of a general wildlife garden within the confines of a tightly 'designed' layout. Although governed by straight lines and geometric curves, the actual design is not a particularly formal one, and adapts well to the concept of the wildlife garden. All the elements of such a garden are present: a colourful nectar border near the house, a large wildlife pond with its complementary wetland, a flowery meadow with a sheltering hedge, and a small area of woodland edge.

The design is one that is derived from the 'pattern' method of approach. It relies on rectangular elements which fit together almost like a puzzle, each containing one or more habitat elements. One objective of this design was the reduction of routine maintenance, and so it does not incorporate any regularly mown grass. Apart from the meadow area, which has a path of timber blocks leading to a log walk-way across the pond, all the other surfaces are either gravelled, as for the eating or barbecue area, or timber decking. A simple pergola near the house gives a vertical element to the garden which echoes the surrounding high walls, and provides support for both climbers and hanging bird-tables.

the trees, and it can meander happily more or less wherever you like. Only when you arrive at a special 'looking place' will it have to widen out a little. This area may need to cater for several observers at once, as one of the nicest things about wildlife gardening is sharing your discoveries with someone else. A bigger area will not only make it easier for you to see but will lessen any wear and tear. Remember to leave the wildlife some totally undisturbed refuges. Even on a small scale a wild habitat can be designed in a way that is visually satisfying. For example, in an garden we often separate or combine plants in a different way from that in which they would grow naturally. Growing a mass of annuals from all over the world in one flower-bed is a good example. In a wildlife garden we would not want to go to this extreme, but it may be an appropriate setting in which to emphasise certain groups of plants. You could plant a far greater concentration of flowering and berrying native shrubs than would occur naturally, or separate out one colour form of a plant and use it to create more impact than the usual mixture.

The clearly defined line of this path (left) gives form to a mass of annuals.

Many herbs are useful to insects, and can easily be incorporated into a front garden such as the one below.

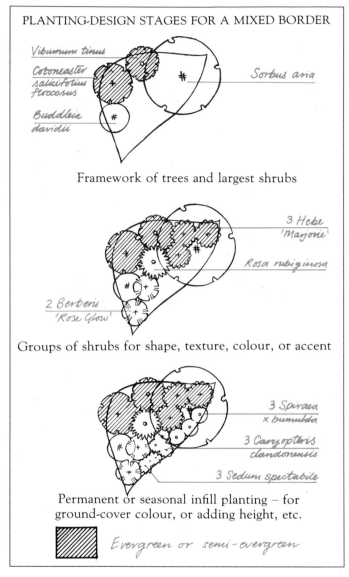

PLANTING-DESIGN STAGES FOR A MIXED BORDER

Viburnum tinus

Cotoneaster salicifolius floccosus

Buddleia davidii

Sorbus aria

Framework of trees and largest shrubs

3 Hebe 'Marjorie'

Rosa rubiginosa

2 Berberis 'Rose Glow'

Groups of shrubs for shape, texture, colour, or accent

3 Spiraea x bumalda

3 Caryopteris clandonensis

3 Sedum spectabile

Permanent or seasonal infill planting – for ground-cover colour, or adding height, etc.

Evergreen or semi-evergreen

choose plants to suit. You will need to ensure that the plants are also suited in terms of soil preference, hardiness, and other such factors. Large areas should be considered in terms of broad groupings of plants rather than individuals, and it is the way that you use these within the different habitats which will create the overall effect. Plant form, colour and texture are just as important as in an ordinary garden because, however attractive the wildlife, you will want the setting in which you see it to be attractive too. In the same way, you will want to provide for something of interest in your garden all the year round. This is sometimes difficult when using only native flora, and it is justifiable to augment it with other suitable species. The choice of these will depend both on the interest that they can provide and, of course, on the wildlife value that they bring into the garden. For instance, you might add the semi-evergreen *Cotoneaster x wateri* to a group of shrubs for the value of its berries. It is worth remembering that if you provide a focus of interest at a particular season, the attention is diverted away from areas of lesser interest. This is best achieved by grouping similar plants together rather than dotting them about where their impact will be lost. In winter a dull corner can be brightened by adding a variegated or golden evergreen to a group of different sized shrubs, and a winter flowering shrub may scent the air near the back door. In fact, any sweet-smelling plants are appreciated near to the house, and scented climbers are especially good on warm walls which seem to intensify and reflect the perfume. Visiting insects, especially moths on a warm night, will be drawn to honeysuckle and summer jasmine. Plants which flower in early spring are valuable too for untimely butterflies tempted out of hibernation by an unusually warm day.

This needs to be done on a large scale to be effective, and you must first be quite sure that it is really compatible with your main aims. The tendency for more definition in a garden than in a natural situation has the effect of dividing areas into visible entities and making maintenance easier. In fact you will probably do this without thinking – for example, making a clean edge to a pond which can then be reached more easily, whereas the natural pond may well have a wide margin of black and sticky mud or impenetrable reed-bed.

What to plant

Now is the time to look at the broad planting categories that you have decided upon, and work out what they will consist of. The way to do this is to start with the requirements, such as dense shrub cover, and then

This carpet of thyme will be especially attractive to bees when in flower, but meanwhile gains extra impact from the use of strong foliage contrasts.

Rosebay willowherb

CULTIVATED WEEDS

Many gardeners equate wild plants with weeds, and fear that they will become over-run with them. In fact, we already use native plants in the garden, often without realising it. Jacob's-ladder, pasqueflower, primrose and lily-of-the-valley are just a few of our native garden plants. Sometimes, they are bigger and more vigorous in the garden, because of the higher soil fertility and lack of competition. Other native plants have been replaced by their garden cultivars. They may be bigger and showier, but they may have less scent and less nectar which is also more difficult for the insect to reach. Some native plants are beautiful, but one would hesitate to give them room in a flower border. Rosebay willowherb is one of these; it spreads rapidly underground and, like other rampant plants, needs to be used where it can grow freely and its vigour used to advantage. If you want to grow something like nettles for butterfly larvae, they must be kept under control in some way. A container sunk in the ground is not always enough, and surrounding them by wide paving may be better. Growing them where they can be controlled by mowing or strimming works well if it is frequent enough.

Native attitudes

Some purists would totally exclude anything non-native from the wildlife garden. There is no doubt, however, that non-native plants can be very valuable as an added extra for birds and insects in the wildlife garden. However, it is important that the main bulk of the planting, especially the trees and shrubs, is native. This also applies to special habitats, such as meadows, where the wildlife gardener is trying to establish a balanced and self-perpetuating community. This balance can easily be upset, and the introduction of unusual species, such as a foreign grass, would be unwise. A wetland is rather similar, but it is easier to remove unwelcome mistakes than from a meadow (although not without a certain effort . . .). This means that the non-native plants of most value in the wildlife garden will be shrubs and smaller trees included mainly for birds, and a wide range of herbaceous material in the more 'gardened' areas, mainly for insects.

Clearing the way

The other advice about plants concerns getting rid of those you do *not* want – the *real* weeds. This is important when you are establishing a new habitat, especially if you have an old or neglected garden to clear. However, unless you really cannot see for the brambles, it is a good idea to let a growing season go by before doing anything major. There may be interesting plants, even useful habitats that could be developed,

and resident wildlife that you will not want to throw out on the streets. Identify and mark anything to be protected, and then clear the remaining vegetation. Cultivating the ground may be easier mechanically, although rotavators cut up the roots of perennial weeds and distribute them over a wider area. Dig these out first and burn them. Annual weeds can be covered with a mulch, but this needs to be really thick to work effectively and is best used in conjunction with other methods. Weedkillers cannot be recommended in a wildlife garden, although there *are* times when they solve a difficult problem. A contact herbicide (Paraquat) is all you need for annual weeds, but a glyphosate-based systemic herbicide (Tumbleweed or Roundup) for perennials.They must, of course, be used with care and restraint. Weed seeds remain in the soil, and will be encouraged to germinate by cultivation. They will need to be removed, perhaps repeatedly, before they seed again. Thick black polythene can be used, over a period of some months, as a mulch to clear the ground. The best results are obtained if it is left down for a whole growing season, lifting it for a short period in the middle to encourage any further germination of weed seeds. This is a good way to prepare the ground for some of the possibilities covered in the three habitat chapters. It may seem tedious and time-consuming, but it will make establishment and maintenance of your new habitats that much easier when you come to create them.

BRINGING WILDLIFE INTO YOUR GARDEN

Not everyone wants to design their garden solely for wildlife nor, indeed, may they have a significant area to set aside as 'woodland edge' or 'meadow'. We believe that you will achieve the best balance and attract the most wildlife if you carefully think out the design of your garden but, even without this, it is perfectly possible to attract wildlife of many forms into an ordinary garden. This can give the owners tremendous pleasure, and can do a great deal for the wildlife itself, providing both feeding and breeding areas, so it is undoubtedly worthwhile.

It is far less effort to feed the birds, plant a *Buddleia* or put up a bat-box. Such gestures can certainly help wildlife in one form or another – to enlarge its populations, survive a hard winter or migrate more successfully. There is also the advantage that you can 'target' a particular area of interest, or even one favourite species, rather than simply bringing more wildlife of every kind into the garden. For convenience, therefore, we have separated out the various groups that people might particularly wish to encourage in their gardens. However, all living creatures interact to some extent, so you will be making a more valuable contribution to wildlife and conservation if you introduce at least one additional wildlife feature into your garden.

Sights like this newly-fledged chaffinch (opposite) should become
common in your wildlife garden.

Insects and flowers

Insects and flowers are naturally interconnected. They have evolved together, so that each depends on the other. Apart from the flower varieties that have been bred by man for their particular features, almost the whole of the incredible beauty and variety of the plant world has evolved as a means of attracting insects. The great majority of plants are pollinated by insects; all the colour, scent and nectar production of flowers is geared to attracting insects, and then ensuring that they visit further flowers of the same species carrying pollen with them. Consequently, if you grow a large variety of flowers, many different insects will visit your garden.

Attracting insects

Insects are a vital part of the ecology of the garden; they eat plants and in turn are eaten by birds, mammals and each other, while plants depend on them for pollination. Their value in assisting birds to survive and breed is inestimable, since almost all birds eat insects at some time, and all garden birds feed their young wholly or almost completely on a diet of insects. If you suffer from mosquitoes, cabbage whites or wasps, you may feel that you have enough insects already; but, in fact,

INSECTS ALL OVER

There are extraordinary numbers of insects, as well as the numerous other invertebrates such as spiders, worms and snails. In Europe as a whole, there are over 100,000 species of insects, while Britain alone has well over 20,000. Therefore you are never likely to identify more than a small proportion of them, and most people concentrate on the butterflies, dragonflies and a few other large species. It also means that there is an insect (or more likely a dozen) for every niche and, whatever you do to diversify your garden, you are almost certain to attract more insects, and most should be welcomed whether you can recognise them or not.

Newly-hatched common green bug nymphs next to their egg-cases.

the great majority of insects are harmless or beautiful and often both, and encouraging a greater number and variety of them can only do good.

There are a number of specific, not very difficult ways to attract a range of interesting, attractive and most often beneficial insects into any garden. Apart from the beauty they bring, they will improve the setting of seeds and fruit through better pollination. You are also likely to increase your bird population by producing more food for them, and may tempt some less common species like the spotted flycatcher to visit. You will also contribute to the vast 'aerial plankton' of invertebrates that rises above our towns and cities, to

be fed on by swifts, martins, bats and other specialist insect-feeders. It is reckoned that a single brood of swifts, for example, is fed about 20,000 insects and other invertebrates *per day*, so the total numbers consumed by all these aerial insect-eaters through the summer is phenomenal. While you obviously cannot relate the insect production from your garden to any specific family of swifts – they travel huge distances in search of insects and London-based birds have been recorded gathering *en masse* to feed over The Wash – you can at least feel that you are contributing to the survival of more of these wonderful creatures.

Insects have complex life-cycles involving many

An example of a simple, yet effective and attractive, border to attract insects close to the house. This takes up very little space, yet the Buddleia bush attracts masses of butterflies, and the marguerites and other flowers attract a range of nectar-feeding insects.

stages, and to encourage them to stay in the garden you need to cater for all these stages. However, we really know very little about the precise requirements of even the most common insects, like butterflies, and next to nothing about the needs of the majority. So it becomes more a matter of providing situations that are known to be well liked by a range of species together with a few extra ideas designed to attract specific insects, such as common butterflies or dragonflies.

· PLANTS RECOMMENDED AS NECTAR SOURCES FOR INSECTS ·

Alyssum *A. maritimum* or *A. alyssoides*
Amelanchier canadensis
*Apple mint *Mentha* × *rotundifolia*
Aubretia (**B**)
Bird cherry *Prunus padus* (**N**)
Blackthorn *Prunus spinosus* (**N**)
*Bramble *Rubus fruticosus* (**N**)
Buddleia davidii in most forms (**B**)
*Candytuft *Iberis gibraltarica*
Caryopteris × *clandonensis*
Ceanothus spp. (**B**)
Colt's-foot *Tussilago farfara* (**N**)
Convolvulus tricolor 'Blue ensign'
*Corncockle *Agrostemma githago* (**N**)
Cornflower *Centaurea cyanus* (**N**)
Cuckoo flower *Cardamine pratensis* (**N**)
*Dame's-violet *Hesperis matronalis* (**N**)
Daphne odora
*Devil's-bit scabious *Succisa pratensis* (**N**)
Erysimum linifolium
Field scabious *Knautia arvensis* (**B**) (**N**)
Fleabane *Pulicaria dysenterica* (**N**)
Forget-me-nots *Myosotis spp.* (**B**) (**N**)
Globe thistle *Echinops ritro* (**B**)

Goldenrod *Solidago virgaurea* (**N**) and *S. canadensis*
Greater knapweed *Centaurea scabiosa* (**B**) (**N**)
Hawthorn *Crataegus monogyna* (**N**)
*Hebe spp., including *H.* × *andersonii variegata*, *H. albicans*, *H. brachysiphon*, *H. salicifolia* (All **B**)
Helichrysum
*Heliotrope *Heliotropium* × *hybridum*
*Hemp agrimony *Eupatorium cannabinum* (**N**) and *E. purpureum*
Honeysuckle *Lonicera periclymenum* (**N**) and others
*Hyssop *Hyssopus officinalis*
*Ice plant *Sedum spectabile* but **not** 'Brilliant' or 'Autumn Joy'
Jasione perennis
Knapweed *Centaurea nigra* (**N**)
Lavender 'Dwarf munstead blue' (**B**)
Lilac *Syringa* sp.
Ligularia clivorum
Limnanthus douglasii (**B**)
Marigolds, of all sorts (**B**)
*Marjoram *Origanum vulgare* (**N**)
*Michaelmas daisies *Aster novae-belgii* and others
Mignonette *Reseda odorata*

pear, Wild *Pyrus communis* (**N**)
Petunia
Phuopsis stylosa
Primula and *Polyanthus spp.* (**B**)
Purple-loosestrife *Lythrum salicaria* (**N**)
Raspberry *Rubus idaeus* (**N**)
Sainfoin *Onobrychis viciifolia* (**N**)
Saw-wort *Serratula tinctoria* (**N**)
Soapwort *Saponaria officinalis*
Spiraea × *bumalda*
Sweet scabious *Scabiosa atropurpurea*
Sweet William *Dianthus barbatus*
Teasel *Dipsacus sylvestris* (**B**) (**N**)
Thyme *Thymus drucei* (**B**) (**N**)
Verbena bonariensis
V. venosa
Viburnum bodnantense
V. tinus
*Water dropworts, especially *Oenanthe crocata* (**N**)

Those marked * are especially recommended. Those marked (**N**) are native plants, while those marked (**B**) are particularly good for bumblebees.

The butterfly garden

The most popular and obvious activity is that of attracting adult butterflies into the garden. Most butterflies are highly mobile, and are likely to find their way into almost any garden, and will stay if there is something to keep them. What you, as a gardener, need to do is plant a selection of suitable butterfly-attracting flowers which are phased so that they provide food for every species, from the earliest hibernators emerging on warm late winter days through to the last small tortoiseshells and red admirals in autumn. There are two excellent books solely on the subject of creating a butterfly garden by Matthew Oates and Miriam Rothschild respectively (see Further Reading on page 151), and these are recommended to anyone wanting to specialise in attracting butterflies.

In this section, we give a summary of the plants which attract butterflies throughout the season, and suggest how to make best use of them in the ordinary garden. As with other forms of wildlife you are likely to attract more, and encourage them to stay, if you design your garden on the principles suggested in the other chapters. But we concentrate here on how to bring butterflies, and later, other insects, into any garden by selecting the right plants.

Butterflies, along with many other insects such as hoverflies, honey bees and bumble bees, need nectar during their adult life for energy to mate, lay eggs, and disperse. They do not need to grow, since this was completed during the larval stage, so their main food requirement is the sugar-rich nectar of flowers, honeydew from aphids or the sugar from over-ripe fruit. Their ability to find nectar sources or caterpillar food plants is very remarkable, and there is no doubt that your chances of attracting reasonable numbers of butterflies to the garden is greatly increased the more suitable flowers you have to offer, thus creating what has been called a 'nectar trap' or even a 'butterfly pub'! It is not just a matter of providing a variety of different flowers to suit the different species at each time of year, but also of having enough of each to make a visible display with a strong enough scent to attract butterflies from far and wide. You should also remember that butterflies, like most insects, love warmth, shelter and sunshine, so your displays of flowers should be placed in such a suntrap.

PLANTS FOR ATTRACTING BUTTERFLIES

For once, we can leave the realm of 'native species are best', and choose from a great range of both exotic *and* native species – many of which are very beautiful garden flowers in their own right – to produce a continuous supply of nectar through the season, though it is wise to avoid double flowers as they are often sterile. Different insects have different mouthparts which allow them to collect nectar or pollen from many different types of flowers. Some are general feeders visiting open, easy flowers like buttercups, where the nectar can be readily drunk. Others, such as butterflies, have a long proboscis and are able to reach down to the bottom of the corolla tube for nectar. Some flowers, especially white and light-coloured ones, emit their strongest scent – their advert to say that nectar is available – from dusk onwards, and these attract night-flying insects, particularly moths. Although there are native flowering plants that can

fulfil all these requirements, trial and error has shown that many introduced species are as good, or better, but the native flowers are greatly preferable for breeding as the larvae cannot adapt easily to different food material. Nectar is broadly the same all the world over, and European insects have no objections to feeding on Brazilian, Chinese or Australian nectar!

The choice of plant species is up to you, and will depend on soil type, latitude, local availability and your own preferences. If you want them to fit in with a scented garden, a silver-foliage border or all blue display, then you can choose accordingly. The list opposite is not exhaustive and new varieties are always appearing, so if you see a plant in a nursery being visited by butterflies, buy it! In our experience, the advice on butterfly plants given by garden centres and non-specialised nurseries is not particularly good, and is often based on folklore, the wrong varieties of otherwise good species, or, even occasionally, the plants they have most stock of, though some nurserymen are very knowledgeable on the subject.

An idealised butterfly border that makes use of plants that will continue to provide nectar and colour throughout the season (adapted from a design by Mathew Oates). It is important that the border is well-sheltered.

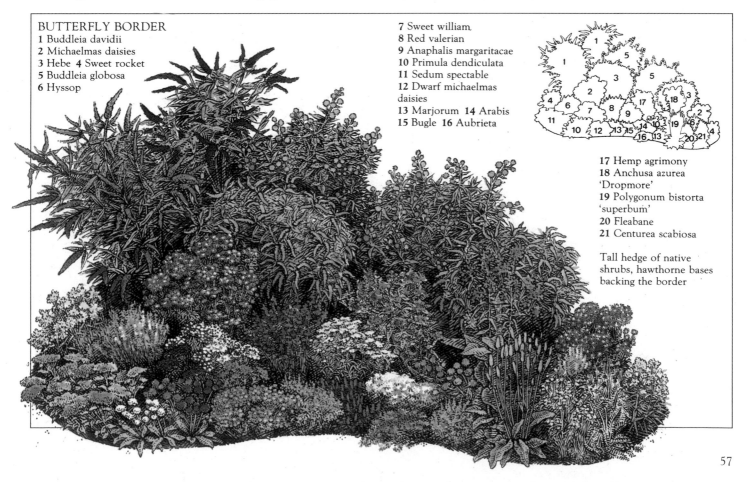

BUTTERFLY BORDER
1 Buddleia davidii
2 Michaelmas daisies
3 Hebe 4 Sweet rocket
5 Buddleia globosa
6 Hyssop

7 Sweet william
8 Red valerian
9 Anaphalis margaritacae
10 Primula dendiculata
11 Sedum spectable
12 Dwarf michaelmas daisies
13 Marjorum 14 Arabis
15 Bugle 16 Aubrieta

17 Hemp agrimony
18 Anchusa azurea 'Dropmore'
19 Polygonum bistorta 'superbum'
20 Fleabane
21 Centurea scabiosa

Tall hedge of native shrubs, hawthorne bases backing the border

---

THE BUTTERFLY BUSH

No butterfly garden is complete without the 'butterfly bush' Buddleia davidii, *in its many forms, or one of the other* Buddleia *species. They are undoubtedly one of the best means of attracting butterflies into your garden especially if carefully managed (though not all species will feed on them). One writer records seeing over 70 individuals of 12 species on one bush! If you plant several bushes and prune some earlier in the spring and others later, while leaving one unpruned, you can extend the flowering season. Also by cutting off the main inflorescences as they fade, you will encourage the lateral branches to carry on producing flowers. A bush planted in a sheltered corner against a wall and allowed to grow, will produce a small tree which flowers at upstairs window level in a way which is rather different.*

A comma butterfly feeding on the variegated Buddleia 'Harlequin' in a garden.

The choice of high summer species is not especially important as there is a wealth of plants which flower from June to September, but it may be useful to mention a few of the early and late-flowering species to look out for here.

SPRING

Spring is a difficult time for butterflies as they emerge from hibernation to seek nectar sources, especially if they wake very early in the year. There is a limited range of species on offer though you can at least provide something to be ready for them. *Aubretia* is a popular rock garden plant which thrives in dry sites. Although not a particularly good butterfly plant, its early flowers will attract some species at a time when little else is about, especially if grown somewhere warm and sheltered; it is also good for early hoverflies, bee-flies and other spring species. Bugle *Ajuga reptans* is a most attractive, free-flowering plant which spreads readily in slightly damp places, and is one of the most popular early plants with butterflies. The native species is generally better than the coloured-leaved varieties which are grown more for their foliage and ground-covering abilities, though some of these can be good. Honesty *Lunaria annuum* is familiar in gardens, grown mainly for its attractive fruits, but its purplish flowers are visited by butterflies and other insects; and orange-tip butterflies will breed on its foliage. The wild primrose *Primula vulgaris*, which can start flowering extremely early in sheltered places, is a good flower for several early butterflies, including orange-tips, and it is much loved by such other early insects as the bee-flies (*Bombylius spp.*). The spurge-laurels or *Daphne spp.*,

---

Honesty (Lunaria biennis) *makes an excellent plant for both breeding and feeding butterflies its variegated form is an attractive garden plant.*

Brimstones, such as this male shown here visiting knapweed, may visit gardens in early spring after hibernation, or late in the summer.

such as *Daphne odora*, are very early-flowering, strongly scented and attractive, and will probably be visited by anything that is about. The beautiful wild cherry *Prunus avium*, which is rarely planted in its native form, perhaps because it ultimately grows too big, is a good source of nectar for many 'generalist' feeders, and it also attracts over-wintered peacock butterflies. If you have space to let the tree grow, you will find it a haven for most other forms of wildlife, too.

Other excellent spring flowers for butterflies include blackthorn, which can be grown as part of a hedge or thicket to give shelter to other areas; willows or sallows, such as *Salix capraea*, the early 'pussy willow' catkins of which provide food for a few butterflies and many more moths and other insects; and even dandelions, which would probably be popular garden flowers if they were not both native and invasive, are quite good for butterflies.

AUTUMN

The other time of year that demands a little thought is the period from September until the last butterfly dies, migrates or goes into hibernation. The majority of garden flowers cease to produce much in the way of nectar-bearing flowers after early September, so more selection is called for. The main butterflies around at

this time of year are the Vanessids, particularly small tortoiseshells, red admirals and commas, though you will also find late brimstones and others. Despite their close relationship to each other, and their habit of feeding together on *Buddleia* in summer, the preferences of the Vanessids diverge in the autumn. Small tortoiseshells love the ice-plant *Sedum spectabile*, which may be covered in literally dozens of these butterflies on warm days, though it is important to avoid the variety 'Autumn joy', which has little to offer. Michaelmas daisies, in most of their various forms, are very attractive to autumn butterflies and will produce flowers late into the season. The electric blue shrubby flower *Caryopteris x clandonensis*, which flowers from late summer until well into the autumn, is another good butterfly and general insect plant. An annual, the common heliotrope, *Heliotropium x hybridum*, will also provide nectar bearing flowers until the frosts knock it back. It is a species which is well worth planting, for its summer value too, though it is not particularly easy to obtain or grow. It is also worth remembering that several late butterflies, especially commas and red admirals, love plums and pears.

Persuading butterflies to breed

The adult butterflies which are attracted to these nectar sources may have emerged from pupae considerable distances away, even from another country. There is no guarantee, though, that any of them will stay in your garden to breed unless you make specific provision for them. There are no 'garden butterflies' which are wholly dependent upon gardens and, in fact, relatively few butterflies will breed in the average garden. Nevertheless, it is well worth making a special effort for these. There is nothing more exciting than finding some butterfly caterpillars about to pupate, and then watching them emerge as the adult butterfly all in your own garden.

Persuading butterflies to stay and breed is by no means a simple matter, and it is not just a question of growing their larval (caterpillar) food-plants. Once you realise that quite a number of butterflies lay their eggs on 'grass' (including many of the commoner ones), yet there are numerous grassy places where such butterflies do not occur, you begin to see some of the problems. The more we discover about butterflies, the more subtle their requirements are found to be, and re-creating the right conditions is not always easy.

THE NETTLE PATCH
Apart from the two common 'cabbage whites', the large and small white, which hardly need attracting, the easiest group of butterflies to encourage are probably

The mint beetle is not always popular with gardeners, though it is one of the most striking of small insects, and rarely does serious damage.

the nettle-feeders. Stinging nettles are the caterpillar food-plants for commas, small tortoiseshells, peacocks and red admirals – four of the commonest and most welcome garden visitors. The trouble is that stinging nettles are hardly the most welcome of garden plants, and they have a nasty habit of spreading rapidly. If you do want these butterflies to breed, you have to site your nettles carefully and make sure they are isolated from the more precious parts of the garden by, for example, a paved path or by growing them in an old tub buried in the soil. They need to be in a warm sheltered place, preferably against a wall or fence and the small tortoiseshell in particular prefers young nettle growth. It is therefore best to cut down at least a part of your nettle patch in late June or early July (checking for

---

IVY

The common wild ivy Hedera helix is an excellent autumn nectar source. In general, the cultivated varieties do not match the wild species since they produce more foliage than flowers, but wild ivy can be a most attractive plant en masse when allowed to clamber freely. It is one of the best all-purpose wildlife plants, and we shall mention it again. Barely touched by frosts, it produces nectar from its heavily, scented greenish flowers and, in late autumn when little else is in flower, attracts red admirals, commas, and painted ladies. For the small garden, it is best grown as the adult form, which you rarely see for sale, and can be produced from a cutting taken from the flowering part of the ivy; this makes a compact bush, without the extensive creeping juvenile foliage. Ivy is also an important larval food-plant for the pretty holly blue butterfly, one of the few blues likely to visit gardens. The ivy flowers will also attract insects and it also makes excellent roosting and breeding cover for birds, a good hibernation site for many animals, including butterflies, and its black berries are eaten, albeit patchily, by birds in winter.

resident caterpillars first) to allow the next generation of butterflies to use the regrowth. If you look at nettles in rough grassy areas that have been cut back, by July they are often well covered with small tortoiseshell caterpillars. You do not need a particularly large clump of nettles, and in some respects a small patch may be better.

Butterflies to look out for

THE HOLLY BLUE

Holly blue butterflies are regular visitors to gardens and often breed, though relatively few people seem to notice them. If you see a blue butterfly flying high over the shrubbery in spring, it is almost certain to be a holly blue, but you are unlikely to be able to catch up with it to confirm your identification. They do not come to nectar as often as other species, though their two preferred larval food-plants, holly and ivy, are both common in gardens. In spring, the female lays eggs on and around the flowers of holly and these have to be female for the caterpillars to survive as they eat the developing fruits; unfortunately the butterflies quite often get it wrong. The butterflies that emerge from these larvae, after pupation, lay their eggs on the buds of common ivy, and the larvae feed on the flowers and

Seen close to, the holly blue butterfly (above) is a very beautiful insect, and is surprisingly common in gardens. Its life-cycle is unusual in that the spring and summer generations feed on different food plants, as shown in the diagram below.

HOLLY BLUE LIFE-CYCLE

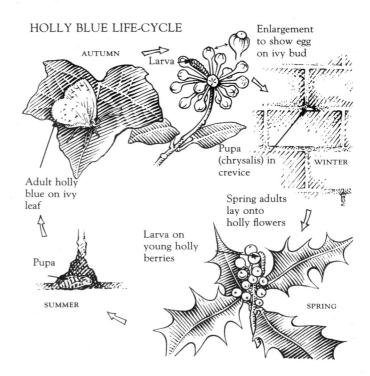

A close-up of the flowers of sweet rocket, or Dames violet, with the pink egg of an orange-tip butterfly laid on the flower on the right.

fruit before spending most of the winter in the pupal state. It is therefore not difficult to provide for this species if you have well-established ivy and fruiting holly, though the butterfly is subject to unaccountable fluctuations in numbers and is not very predictable in its habits.

WELCOME WHITES

Two butterflies which have roughly similar food-plant requirements, and can both be attracted into the garden, are the related orange-tip and green-veined white butterflies. The former is a small spring species, unmistakeable among north European butterflies for the marked orange tips on the male's wings. The green-veined white is an attractive butterfly, which bears an unfortunate resemblance to the infinitely more destructive small white. The green-veined white, distinguishable by its dusting of green and yellow along the veins on the undersides of the wings, has not taken to the cultivated members of the cabbage family so has not become a pest. You can quite easily encourage the butterfly to breed by providing honesty, *Arabis*, horseradish or its preferred native plants – hedge mustard and the cuckooflower.

· LARVAL FOOD-PLANTS FOR SOME · GARDEN BUTTERFLIES AND MOTHS

· BUTTERFLIES ·	
Red admiral	Stinging nettle
Comma	Stinging nettle, hop, currant, gooseberry
Common blue	Bird's-foot-trefoil, small legumes
Holly blue	Holly, ivy
Brimstone	Alder buckthorn, purging buckthorn
Meadow brown	Soft grasses
Small copper	Sorrel, dock
Gatekeeper	Soft grasses
Small heath	Soft grasses
Orange tip	Cuckooflower, garlic mustard, dame's-violet
Painted lady	Thistles, burdock, nettles
Peacock	Stinging nettles
Ringlet	Grasses
Dingy skipper	Bird's-foot-trefoil, horseshoe vetch
Large skipper	Cock's-foot, other grasses
Small skipper	Grasses
Speckled wood	Grasses
Small tortoiseshell	Stinging nettles
Wall	Coarse grasses
Green-veined white	Cuckooflower, dame's-violet, garlic mustard, watercress
· MOTHS ·	
Elephant hawkmoth	Rosebay willowherb, clarkia, fuchsia
Eyed hawkmoth	Apple, willows
Lime hawkmoth	Lime
Poplar hawkmoth	Poplars, willows
Privet hawkmoth	Privet, lilac, ash
Burnet moth	Bird's-foot-trefoil, small legumes
Cinnabar moth	Ragwort, groundsel

The orange-tip uses similar species in the wild, being particularly fond of cuckooflowers, but the larvae eat the developing fruits rather than the leaves. They can also be attracted in to clumps of sweet rocket or honesty though, for some reason, the females prefer to lay on small clumps rather than great masses, perhaps to avoid being heavily parasitised.

Most of the other native butterflies that you might conceivably encourage to breed in gardens have food-plants that do best in carefully controlled habitats such as the meadow or woodland edge; these are described in the appropriate chapters. Such butterflies include the common blue, a number of the 'browns' which feed on long grass, the small copper and possibly the small or large skippers.

Attracting other insects

Butterflies are the most attractive insects to encourage into the garden, but there are thousands of other interesting and beneficial species, which should all be welcomed. Moths, including the extraordinary hawkmoths, dragonflies and damselflies, grasshoppers, bush-crickets, stag beetles, longhorn beetles, ground beetles, hoverflies and ladybirds are just a few of the better known ones.

If you adopt the planned wildlife garden approach outlined throughout this book, you will inevitably get more insects of all sorts, but you can also do a little extra for the other groups, even in the simplest garden

given over largely to other purposes. Insects are so varied in their requirements, that they will be present whatever you do, and anything which increases the diversity of flowers and the mosaic of situations in the garden will inevitably bring in more. There are also some specific flowers that you can plant to attract certain species or groups of insects.

HOVERFLIES — AN ECOLOGICAL APHICIDE

Hoverflies are attractive small insects, with hundreds of different species in Britain alone. Most species that regularly visit gardens have larvae which eat aphids (greenfly) voraciously. It has been shown that one hoverfly larva consumes about 600 aphids during its development into an adult – a great ally to any gardener! The females of the aphid-eating types seek out aphid colonies in which to lay their eggs, usually singly, though often up to several hundred in all. The larvae then actively hunt the aphids, peering into likely crannies and even scenting them out. You would hardly wish to provide aphids so that you can attract hoverflies but, you can provide the flowers liked by adult hoverflies, bringing more of them into the garden and increasing the chances of them laying eggs. This is

Some adult hoverflies, such as this Episyrphus balteatus, *bear a passing resemblance to wasps. They are in fact perfectly harmless, and the aphid-eating larvae are definite 'gardeners friends' (right).*

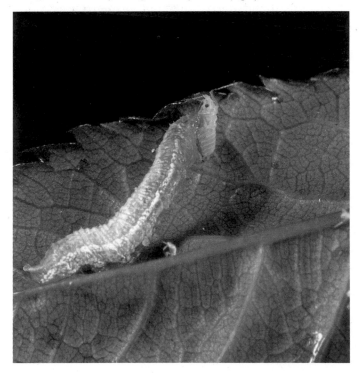

A lime hawkmoth resting on a post at night. If there are lime trees in the vicinity, these beautiful insects are not uncommon as nocturnal visitors.

one of the reasons for planting marigolds and other plants among the vegetables. Hoverflies like a great variety of flowers but tend to prefer broad open flowers or flat-topped groups of flower heads with readily accessible nectar. Some trials by one organisation showed that the poached-egg plant *Limnanthes douglasii* was the best all-round species for hoverflies; it was also good for honey bees. It is a hardy annual and very easy to grow. Second best, in the trials, was the annual blue convolvulus *Convolvulus tricolor* Blue Ensign. Other good plants include nasturtiums, evening primroses, poppies and members of the carrot family like hogweed, Angelica and carrot itself.

NIGHT VISITORS

Moths hardly fulfil quite the same pest control function as hoverflies, but they are beautiful and largely harmless insects. You are unlikely to know much about your garden moths as most of them are nocturnal, and you just see the few that beat themselves against the window or come into the house. It is quite likely though, that you have at least 100 species of moth in or around your garden, and you could have more. Many of the listed butterfly plants will encourage moths, but there are also some excellent plants that are particularly attractive to them. Honeysuckle *Lonicera periclymenum*, especially the ordinary native form, attracts moths from dusk onwards, and you will probably have noticed how its scent is strongest on warm evenings, just when moths emerge. Other good moth nectar plants include the *Nicotianas*, or tobacco plants, which are both decorative and easily grown, the evening primroses *Oenothera spp.*, summer jasmine,

night-scented stock and red valerian. These are all good plants for insects in general. The wild clematis or old man's beard, is also good for moths, though it does not normally find favour as a special garden plant. The tall purple spikes of rosebay willowherb, though thought of as a weed, are beautiful plants and, as well as their useful flowers, their foliage is the food for the extraordinary large elephant hawkmoth. The larvae of this moth grow to a remarkable size, causing some confusion over their identity, and the adult moths are beautiful soft green and red creatures. Alternative food-plants include some of the less invasive willowherbs, like *Epilobium fleischeri* or *E. dodonaei*, or the closely-related *Clarkias*. If you particularly like hawkmoths, it is a relatively easy matter to look up the food-plants of the species that occur in your area, and make sure that you grow them, though many occur in gardens anyway, for example privet for privet hawkmoths, or apple trees for eyed hawkmoths.

The special plants for butterflies, moths and hoverflies, as well as all the other flowers you grow will cater for most other flower-visiting insects. However, there are a few other plants that seem to attract particular insects. For example, we always get the attractive solitary bee *Anthidium maculatum* visiting our 'Lamb's

NIGHT-SCENTED PLANTS
· TO ATTRACT MOTHS ·

Bladder campion *Silene vulgaris* (**N**)
Californian poppy *Eschscholtzia californica*
Dame's-violet *Hesperis matronalis*
Evening primrose *Oenothera spp.*
Everlasting pea *Lathyrus sylvestris* (**N**)
Honeysuckle *Lonicera periclymenum* (**N**) and some
  cultivated spp.
Tobacco plant *Nicotiana spp.*
Night-scented catchfly *Silene noctiflora* (**N**)
Night-scented stock *Mathiola longipetala bicornis*
Petunia × *hybrida*
Phlox, various
Red valerian *Centranthus ruber*
Soapwort *Saponaria officinalis*
Summer jasmine *Jasminum officinale*
Traveller's joy *Clematis vitalba* (**N**)
Verbena *V. bonariensis*, *V. venosa*
White campion *Silene alba*

Those marked (**N**) are native plants.

Lawns don't have to be all grass – this one is dominated by masses of the beautiful hoary plantain, normally a plant of chalk downland.

ear' *Stachys lanata* in the garden, while the flowers of wild-type roses (those with single petals and a mass of stamens in the middle) are extremely attractive to longhorn beetles and other pollen (rather than nectar) feeders. The leaves of any roses are likely to prove an irresistible attraction to the female leaf-cutter bee, which cuts neat, almost circular holes and carries each piece away to make nests for her larvae. This is a remarkable process to watch, as she cuts the circle out in about 10 seconds, before flying off with it tucked between her legs! We have seen these delightful bees listed as pests because they make holes in rose leaves, but surely this is something we can all live with?

A LIVING LAWN

One of the best ways that you can liven up your garden for all insect visitors is to be a little less rigorous about your lawn. The extreme of this is the 'flowery meadow' described on page 112 onwards, but at the other end of the scale you can still make a contribution while maintaining a lawn that fulfils all its normal purposes. If you give up the use of fertilisers and weedkillers, but continue to mow regularly at a slightly higher setting, you will gradually acquire daisies, slender speedwell, celandines, clover, medick, plantains and other low-growing plants. These put on an excellent display and

there are fewer nicer spring sights than a lawn covered with daisies and deep blue speedwell flowers. But they have other effects, too: they make the sward less homogenous, which encourages solitary (harmless) bees to nest; the leaves of the different plants support various harmless herbivorous insects; while in the unseen layers below the surface, the variety, and probably density, of invertebrates will be increased. Naturally, this makes for an excellent feeding ground for many visiting birds, including blackbirds, thrushes, starlings and even the occasional green woodpecker. Unless you acquire excessive amounts of less desirable plants, like stemless thistle or bur medick, then you can use the lawn for most of its normal purposes. It is just a matter of looking at it slightly differently and ignoring the neighbours' comments or the blandishments of advertisers out to sell you all-in-one fertilisers and weedkillers! (page 115 onwards gives more details of managing the lawn in different ways.)

HOLES AS HOMES

It might seem a little odd at first, but you can also provide suitable egg-laying sites for various insects to use. Some species are limited in their distribution by such sites. Many solitary bees and wasps, for example, both of which are well worth having in the garden as pollinators and caterpillar-eaters respectively (they are *not* the ones that come and pester you for fruit or jam!) require holes to nest in. In nature these include hollow stems, holes in trees and rotting wood and cracks in stones. One excellent idea we have seen used in Germany (see picture) involves using a box rather like a roofed bird-table, erected on a post and stuffed with hollow canes and straws of all different sizes. You can also drill holes (of about 5–10mm in diameter) in posts or masonry, and see what uses them – something is bound to although perhaps not what you expected!

LIFE IN DEAD WOOD

Quite a number of interesting insects spend their larval stages in dead wood of one sort or another. As you might expect, dead wood is not exceptionally nutritious, so most need to remain as larvae for several years gradually growing and developing until large enough to pupate and turn into adults. This poses a problem, especially in our increasingly tidy countryside, since

A bundle of hollow sticks attract harmless and attractive solitary bees and wasps to nest in your garden.

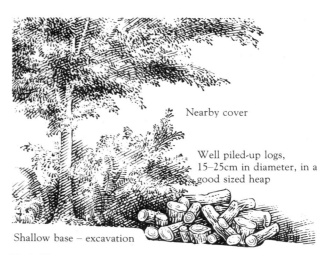

Nearby cover

Well piled-up logs, 15–25cm in diameter, in a good sized heap

Shallow base – excavation

Shaded log-piles, on a dug-out base, are the most successful for attracting wildlife. Standing water however, should not be present; you should not excavate the base if including a hedgehog house.

suitable dead wood is often not present for long enough to allow them to complete their larval stages. Such insects as stag beetles, lesser stag beetles, many of the beautiful long horn beetles, some hoverflies, solitary bees and various associated predators and parasites, all need dead wood as food for their larval stages, and since this is becoming rarer in the countryside, it makes sense to provide it in your garden. Many of the insects visit flowers for nectar in their adult stages, so you will have the pleasure of seeing them as well as helping with their conservation. It is not, however, quite as simple as it is often portrayed as most such insects and other invertebrates have very specific requirements. Some like a particular type of tree, some like dry wood, others prefer it damp. Some need large trees, others need high branches that are dying. Obviously you cannot provide more than a small proportion in an ordinary garden, but a carefully thought out wood-pile for the purpose, together with a few other ideas, should help.

Most dead-wood species fare badly if in full sunshine, and so this is best sited somewhere that is shady for most of the day. Treated wood should be avoided and very small branches have relatively little value. If you can put a mixture of types of wood, for example beech, oak, ash, elm and possibly pine, you will attract a greater range of species. It is best if the bark remains on, at least at first. Thus you should end up with a pile of wood of varying sizes and types, placed somewhere reasonably shady. Our main one is under the apple tree by the hedge. You may feel that you would rather be burning the wood but you are only sacrificing a few evenings' fires for a logpile which will last for several years – and you could always add some of those large pieces which were impossible to split! You will not see much of what goes on in your 'conservation wood pile' but it will undoubtedly be used, and you may be rewarded in due course by the sight, one spring, of male stag beetles or large black and yellow *Strangalia* beetles feeding in your rose flowers. Other animals will probably use the place, too, and you can also put the hedgehog house under the pile.

Most insects not attracted by these suggestions are probably either not suited to gardens, or need very special conditions. Where appropriate, these are covered in the separate 'habitat-creation' chapters. For example, the only way to attract breeding dragonflies and damselflies, is to provide an area of still or slow-moving water such as a pond (see page 128). Although some species may visit your garden to feed briefly, they will not stay unless there is water. Similarly, most grasshoppers and some bush-crickets need an area of rough grass, of varying heights according to the species, and some of these can be catered for by creating the flowery meadows discussed on page 112.

A beautiful male stag beetle, one of the most attractive and striking of the insects which can be attracted into the garden by careful planning.

Insects and pesticides

One of the most difficult concerns facing the potential wildlife gardener is 'to spray, or not to spray'. On the one hand, the possibility of clean, unblemished fruit and vegetables without slugs, aphids or caterpillars, beckons; while on the other, the thought that it would be better to eat fruit and vegetables grown without pesticides and the possibility that it may be killing attractive or harmless animals, weigh against this. These appear to be simple issues, but in reality it is far more complicated. There is a good deal of evidence that most insecticides kill off the vulnerable predators more easily than the target pest species; because the herbivorous species, which are always the pests, can reproduce more quickly than predators, the few which survive the spraying simply flourish in a predator-free environment rapidly regaining their original numbers. The worrying thing is that it forces you to spray again if you want to maintain the *status quo*. Also of concern are the long-term effects of spraying, especially with persistent pesticides, which can be so far-reaching and insidious. You may still see dragonflies coming to your pond, but what if their breeding success rate has been greatly lowered by the toxins they have picked up via their

insect prey? So many species in the countryside have gradually declined (and we are often not sure whether it has been due to habitat loss, pollution, pesticides or some other cause) that it makes sense not to contribute to their problems in your garden. It is easy to be reassured about pesticides because you cannot see them or their effects, but the damage caused by the earlier persistent types, some of which are still in use, was shown to be massive and far-reaching.

There are alternatives to traditional spraying which can be effective in certain circumstances. We have selected some of the better ideas, but anyone wishing to take the concept further should consult the organic gardening publications listed in the Further Reading or contact one of the organic gardening organisations.

SLUGS AND THEIR FRIENDS

Slugs are an undoubted problem. However, they are a prime food for hedgehogs, ground beetles and other predators, so it makes sense to encourage the latter. Slug bait is poisonous to other animals and there is some evidence that it will affect animals eating dying slugs, though possibly not seriously. There are certain considerations of design that will reduce slug damage (see page 41) and you can also try other control

methods such as beer traps or 'slug pubs', or organic slug traps. You can also try putting down a sheet of heavy opaque polythene in an area of high damage, which the slugs will use as a daytime hide-out – you can then simply remove them all at once. With particularly vulnerable plants like lettuce seedlings, you need to give added protection through the worst period. None of these methods will totally solve the problem but, if applied in a concerted way, they will greatly reduce it.

You can try other ideas for different specific pests. Grease bands around the trunks of fruit trees should stop the female winter moths (which are wingless) from climbing the trees, while pheromone traps are available to attract male codling moths. A bacterial spray *Bacillus thuringiensis* is available for use directly on caterpillars, and it is said to affect nothing else. Some doubt has been expressed as to whether its effects may be wider than believed, but this is unresolved. There are other methods of biological control available, too.

Finally, if you do use sprays, be very careful in your choice and follow the maker's instructions implicitly. Using more than recommended will not give better

Hedgehogs are amongst the most frequent of mammal visitors, consuming considerable quantities of slugs.

A COMPARISON OF THE EFFECTS OF DIFFERENT ORGANIC INSECT ·CONTROL METHODS·

● insects affected – not always to same extent
○ should not harm in correct dosage

| | Pests | | | | | Beneficial insects | | | | |
---	Aphids	Sawfly	Caterpillars (large)	Caterpillars (small)	Raspberry beetle	Ladybirds	Hoverflies	Lacewings	Bees	Anthocorid bugs
Soft soap	●			●		○	○	○	○	○
Derris (spray)	●	●	●	●	●	●	○		○	
Nicotine	●	●	●	●	●	○	○	○	●	○
Quassia	●	●		●		○	○		○	○
Bacillus thuringiensis				●		○	○	○	○	
Elder spray (home-made)	●			●		○	●		○	●

control, even though you may feel it will. Avoid wide-spectrum organochlorine insecticides totally, as they are bound to affect your predators and many harmless insects, too. 'Natural' insecticides like Derris and Pyrethrum have lower toxicity and persistence, but they are not specific – Derris, for example, kills ladybird larvae. See table above for an analysis of the effects of organic pesticides. In many cases, you will find that a solution of soft soap is the best answer. For resident pests, try spraying with these insecticides late in the evening, so that the toxicity has been reduced by the time tomorrow's flower visiting insects arrive.

Wild flowers in the garden

Most aspects of using native wild plants in the garden are covered elsewhere, as means of attracting insects, food for birds or as elements of the special habitats in succeeding chapters. There seems little point to us in trying to use the garden as an area for planting rare and unusual wild flowers. Since they will be living in quite artificial conditions, it will not help their wild populations which have almost invariably been reduced by habitat loss rather than anything else. Also there is a danger that, by creating a market for wild plants, you encourage their destruction in the wild.

Weeds

Weeds are the subject of much debate and dislike. They are perhaps best defined as plants growing where they are not wanted. If you follow this definition, you can reduce the number of weeds in your garden 'at a

stroke' by simply getting to like them. If you leave out the most pernicious weeds, like ground elder, the exotic *Oxalis* species, hedge bindweed, couch grass and a few others, the remainder are reasonably easy to control. If you leave the hairy bitter-cress and the pearlwort on the path, and the duckweed on the pond, your garden will be richer rather than poorer for it. You will have to control them, preventing too much colonisation and swamping, and keeping special areas like the vegetable patch clear, but you can eventually achieve a reasonable balance without too much effort. You will also gain a great many insect species, including the beneficial predators which will use your weeds as shelter, cover or food.

Procumbent pearlwort is an abundant weed on paths and paving, and it does very little real harm, though most gardeners attempt to remove it.

The pesticide-free garden

Is it possible to have a garden that is reasonably productive in the absence of insecticides and other sprays? The answer is a qualified 'yes', but the withdrawal of pesticide use can be a gradual one. There are some useful general points to consider on the way to a pesticide-free garden.

● *Avoid preventative spraying or spraying for the sake of it. In the event of a build-up of pests take the line that it is a one-off occurrence, until shown to be otherwise. In other words, delay for at least a year to see what transpires before reaching for the spray. You will know what to look out for the following year and be ready to do something early on.*

● *Most of the garden should simply not be sprayed at all, and this should certainly include any areas set aside for wildlife.*

● *For certain specific problems, such as cabbage white larvae on the Brassica crops, try hand control. If you are in the garden regularly, look for eggs and young larvae and remove them as early as possible; also remove the larger caterpillars whenever you find them.*

● *Try to accept a slightly different standard of fruit and vegetables; holes, marks and scabs may look unsightly, but they will do you no harm, whereas blemish-free fruits which have been sprayed could be slowly poisoning you!*

● *Plant resistant varieties wherever you can, if you are aware of a specific pest that is a problem in your area. For example, raspberries such as 'Malling Delight' are partially resistant to aphids, while some potatoes, such as 'Maris Piper', are more resistant to eelworm.*

● *Try to encourage predators as much as possible. These will naturally do better in a pesticide-free environment, but they will also flourish in a more varied garden. Companion planting, putting selected flowers between the vegetables, breaks up the scent of the vegetables and encourages predators like hoverflies; can be effective, though opinions differ as to whether it really works. It certainly looks attractive and the extra effort involved in harvesting is not great.*

● *If you do need to spray, choose the least damaging sprays, and make sure they are as specific as possible to what you want to control, so that you do not affect other species unduly (see table left for some recommendations).*

· SOME CORNFIELD WEEDS ·

If suitability is limited by soil type, preferences are indicated by the following letters:
N neutral
A acid
C alkaline, calcareous
M moist or wet
D dry, well drained

	Soil preferences	
Corn buttercup *Ranunculus arvensis*	N–A	M
Corncockle *Agrostemma githago*		
Cornflower *Centaurea cyanus*	N–A	D
Corn marigold *Chrysanthemum segetum*	N–A	
Common poppy *Papaver rhoeas*		D
Flax *Linum usitatissimum*	C	D
Ground pine *Ajuga chamaepitys*	C	
Night-flowering catchfly *Silene noctiflora*	C	
Pheasant's-eye *Adonis annua*	C	D
Scentless mayweed *Tripleurospermum inodorum*		
Venus's-looking-glass *Legousia hybrida*	N–A	
Wild pansy *Viola tricolor*		M
Barley		
Bearded wheat		
Oats (not wild)		

CORNFIELD FLOWERS

One interesting and attractive idea, which can certainly do no harm and may be useful, is to set aside a small area of the garden as a 'cornfield weed plot'. Most of our traditional arable weeds, like cornflower, corncockle venus's looking grass, and thorow-wax, have declined to the point of extinction through improved seed-cleaning, more herbicides and altered timing of cropping. Those with short-lived seeds have virtually gone, while those with long-lived seeds, like poppy and charlock, will always survive to take advantage of any opportunity.

If you set aside a small area – and only a few square metres are necessary – which you cultivate each year and sow with a mixture of crop (e.g. wheat, or oats) and cornfield weeds, together with some other varieties, you will end up with a highly attractive mixture of flowers that mimics the old-fashioned cornfield which you can no longer see in the countryside. The ways of achieving this effect are described in more details on page 112.

A crop of oats sown with a mixture of arable weeds, including corncockles, common corn poppies, and flax, makes an attractive 'cornfield-edge'.

Attracting birds, mammals and

Birds are undoubtedly the most popular of wild garden visitors. They are all beautiful and most are harmless or beneficial in some way. Almost all are active during the day, unlike many mammals, so they tend to be more noticeable. There are many ways in which you can attract birds into the garden and persuade them to stay, from the simple expedients of putting out food through to growing shrubs and trees that may be suitable as nesting or roosting sites.

Food for birds

People have provided food for birds for a very long time, and this is one of the easiest ways to attract them into your garden, particularly in late winter when natural food supplies are at their lowest. Some birds have even changed their eating habits as a result of garden feeding, and it may have helped some species survive the rigours of winter. Blackcaps now regularly stay in Britain throughout the year, and they are often seen at bird-tables having switched from their normal diet of insects which is largely unavailable in winter here. Siskins have become much commoner in some areas, partly due to the extension of suitable conifer plantations as breeding sites, but they also visit gardens regularly from the New Year onwards feeding avidly on nuts in particular.

There are two ways in which you can approach bird feeding. You can either put out food in controlled amounts in specific places, or you can grow the plants

It is particularly important to put food out for birds in winter.

GARDEN VISITORS

For interest, the 'Top 12' birds found coming to gardens to feeders in a recent survey were:

House sparrow	Great tit
Starling	Dunnock
Blue tit	Robin
Chaffinch	Collared dove
Blackbird	Coal tit
Greenfinch	Song thrush

It is unlikely that your own garden would have only these birds but most of these species will probably occur. For comparison, the approximate 'Top 12' birds coming into the author's garden last winter were:

Blue tit	Robin
Great tit	Dunnock
Greenfinch	Starling
House sparrow	Siskin
Song thrush	Coal tit
Blackbird	Blackcap

which birds require or which support the creatures they feed on. Either can be successful, and there is obviously no reason why you cannot do both.

Feeding of birds should be done with some care, as it is possible to feed the wrong food at certain times. Also the dependence of birds on a specific feeding site can make them very vulnerable to predation by, for example, domestic cats, and they may spend dangerous amounts of time waiting about for food in a cold spell when you have gone to Tenerife for a Christmas break, so take care how you organise your bird-feeding!

other animals

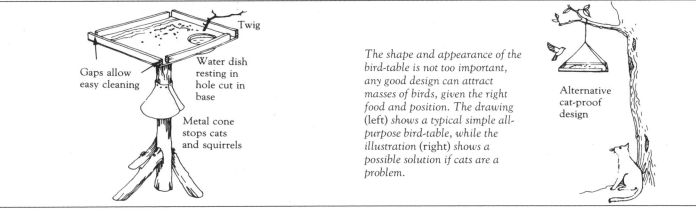

Gaps allow easy cleaning

Twig

Water dish resting in hole cut in base

Metal cone stops cats and squirrels

The shape and appearance of the bird-table is not too important, any good design can attract masses of birds, given the right food and position. The drawing (left) shows a typical simple all-purpose bird-table, while the illustration (right) shows a possible solution if cats are a problem.

Alternative cat-proof design

BIRD-TABLES

Bird-tables need to be carefully sited. Most people want them to be close to the house and easily visible from a well-used window – the birds will soon get used to the movement inside. They should be readily accessible, whatever the weather, so it is usually better to have a hard access to them to preclude crossing wet, muddy grass every winter day. Finally, they should be sited so as to make it difficult for cats to attack the visiting birds. We have found that birds are generally most vulnerable when they are approaching the table, rather than when they are actually on it, though if you put your table close to cat-cover the birds will certainly be attacked. Try to make sure that the table is at least 1.5 metres from any cover and it should certainly not be so low that a cat could jump onto it. Hanging tables are often better than standing ones. If you find that your visiting birds are tending to perch or gather somewhere that they can be easily caught by cats, you may have to think again and either move the table or the gathering-place. Other predators, such as sparrowhawks, are likely to be much less frequent, and should be welcomed as an exciting addition to the garden bird list.

Different species of birds and even, to some extent, different individuals have particular preferences as to where they feed and what they eat. To attract as many birds as possible, you need to be aware of and cater for these needs. The traditional bird-table is really just a starting point, and it should be supplemented by food in hangers of various sorts, on the ground, and jammed into crevices in logs or posts. This not only gives opportunity for more species, but it also helps to give the less aggressive species a chance to find food despite the presence of starlings, black-headed gulls or a grey squirrel. Birds like the hedge sparrow will prefer to feed on the ground, tits and siskins like hangers, while the great spotted woodpecker prefers to winkle his fatty morsels out of a crevice in some old wood, though in times of severe shortage these distinctions may break down. Having several feeding areas is usually better than just having one as it gives more birds a chance.

WHEN TO FEED

It is generally best to phase out feeding in spring, as natural food sources become available and young birds start to hatch, then gradually start again in late autumn as the supply of fruits, seeds and insects dwindles. Throughout the winter, try to maintain a regular time of day to put food out – after breakfast each morning for example, and keep this up if you are getting a regular number of birds. If you are going away, perhaps a neighbour could be persuaded to continue the regime in return for your watering their windowboxes next August, or you can buy hopper feeders which are moderately successful.

THE BIRD-TABLE MENU

The range of food provided should include seeds and fruits – such as sunflower seeds, bird-seed mixtures and raisins – fatty foods which can be based on meat or vegetable fat – these are particularly good energy sources in midwinter. Peanuts are well-known favourites with tits, siskins and many other birds that learn to extract them from the feeders. Robins love little bits of cheese, raisins or, if you feel like going to the trouble of breeding them, mealworms. The usual mixture of bread-crumbs culled from the breadboard, breakfast plates and the bottom of the toaster is a useful extra, especially if it is wholemeal bread, but, if you can, it is better to put out more of a selection of foods.

You can buy ready-made bird-food mixtures in various guises and most of these are good, especially if recommended by the Royal Society for the Protection of Birds or other relevant organisations. Over-ripe fruit, either from your own trees or from the local shop, will be well-liked by blackbirds in particular, but also by other visiting members of the thrush family such as redwings. It will also be found by the various insects which are still around in late autumn. The availability of imported or stored fruit greatly increases the possibilities for using this excellent food source.

A young greater spotted woodpecker feeding at a log from fatty scraps pushed into the crevices.

· PLANTS WHICH PROVIDE FOOD FOR WILD BIRDS ·

Apple, cultivated	Evening primrose *Oenothera biennis* and others	Privet *Ligustrum vulgare* (**N**) – the wild species, uncut
Berberis stenophylla	Fruit bushes, blackcurrant, raspberries	*Pyracantha coccinea*
Bird cherry *Prunus padus* (**N**)	Guelder rose *Viburnum opulus* (**N**) (yellow-fruited varieties are less good)	*Pyracantha* 'Mojave'
Brambles *Rubus fruticosus* (**N**) and others		*Pyracantha* 'Orange glow'
Burdock *Arctium minus* (**N**)	Groundsel *Senecio vulgare* (**N**)	Rowan *Sorbus aucuparia* (**N**)
Chickweed *Stellaria media* (**N**)	Hawthorn *Crataegus monogyna* (**N**)	Red chokeberry *Aromia arbutifolia*
China aster	Hogweed *Heracleum sphondylium* (**N**)	Snapdragon *Antirrhinum* spp.
Cosmos	Holly *Ilex aquifolium* (**N**)	Spindle *Euonymus europaeus* (**N**)
Cotoneaster spp. especially *C. horizontalis*, × *watereri*, *frigida*, *simmonsii*, but **not** *C. conspicuus decorus*	Honeysuckle *Lonicera periclymenum* (**N**) and others	Stinging nettles *Urtica dioica* (**N**)
	Ivy *Hedera helix* (**N**)	Strawberries
Cow Parsley *Anthriscus sylvestris* (**N**)	Knapweed *Centaurea nigra* (**N**) and others	Sunflowers *Helianthus annuus*
Crab Apple *Malus sylvestris* (**N**) or John Downie (not Golden hornet)	Meadowsweet *Filipendula ulmaria* (**N**)	Teasel *Dipsacus sylvestris* (**N**)
Dog rose *Rosa canina* (**N**)	Michaelmas daisies *Aster novae-belgii* and others	Thistles, various. *Cirsium* (**N**), *Carduus* (**N**), *Onopordon* etc.
Dandelion *Taraxacum officinale*		Wayfaring tree *Viburnum lantana* (**N**)
Dogwood *Thelycrania sanguinea* (**N**)	Plantains, e.g. *Plantago media*, *P. major* (**N**)	Woolly thistle *Cirsium eriophorum* (**N**)
Elder *Sambucus nigra* (**N**)		Yew *Taxus baccata* (**N**)
Eleagnus angustifolia		Those marked (**N**) are native plants.

Plants for birds

If you follow any of the advice given in this book, you will inevitably increase the number of insects in the garden, and many of these will provide food for birds at some stage of their life-cycles. You can also grow specific plants which will directly provide food, usually in the late summer or autumn as their fruits mature. Many of these are highly attractive plants as well as bearing nutritious berries.

A list of plants recommended for planting as food for birds is given above. As a general rule, we feel that it is well worth planting ordinary fruit trees and bushes anywhere you have the space for their irresistible combination of flowers and fruit as bird and insect food. Just think of the time most gardeners spend *preventing* birds from eating the fruit! You can protect a proportion of the fruit in a cage or under netting, but leave others for the birds to find. If you think of the apple, plum and other trees as part of your provision for wildlife, you will be delighted rather than upset by the sight of a pair of bullfinches feeding on the blossom.

Soft fruits such as currants, gooseberries and raspberries, which have excellent nectar-rich flowers, are good in summer while blackberries, crab-apples and cooking apples provide food for autumn. Crab-apples come in all different colours and forms, and some of the most attractive are also good as bird food – though,

The combination of ripening black grapes and autumn foliage on this wild-type grape-vine looks very attractive, and the grapes can also prove an irresistible lure for birds, if you leave some unprotected for them.

of course, the display will not last for long. Several varieties of grapevines will grow successfully even quite far north, and are now widely available. They are good for climbing over a shed or wall – in contrast to the heavily-pruned commercial wine forms – and through September and October they produce a beautiful crop of fruit among the bronze and green foliage. These are favourites with blackbirds, starlings and other birds – not to mention children – but there are usually enough grapes left on a well established vine to make your wine or even to eat a few bunches.

With any fruit-bearing tree, whether it is a rowan, apple or blackberry, you can extend the season by harvesting the fruits and drying or freezing them for use later. Some, like rowans, dry quite well in the British climate, but most others do not and freezing is better. Put the individual fruits on trays in the freezer and then bag or box them once they are solid, as they will then remain separate for ease of use, later on.

HOLLIES, THISTLES AND TEASELS

For most of the other plants suggested in the list on page 75, it is simply a matter of making a choice according to your soil, situation and preference; but it is worth adding a word of caution about hollies, in particular. The cultivated forms of holly include many beautiful shrubs with striking foliage and the possibility of carrying abundant berries as bird food. However, nearly all have separate male and female plants and only

Some of the varieties of the native holly can be very beautiful, such as this Ilex aquifolium *Madame Briot', whilst still retaining the ability to produce berries and attract birds.*

MAKING CHANGES IN AN ORDINARY GARDEN

This garden (below) was typical of many gardens everywhere. A piecemeal approach had resulted in an awkward layout, practically and visually. Unimaginative planting, did not enhance the 'bitty' arrangement of the rose and flower beds. The front garden was less cluttered, but not particularly exciting.

A Badly-positioned garden shed
B Coloured sycamores too large, not good for wildlife
C Paved area too dominant
D Ugly rockery with sunken concrete feature
E Dwarf conifers and standard roses compete for attention

The aim was to improve the basic design of the garden without too much expense or upheaval, and to encourage wildlife at the same time. The final design is uncomplicated, as befits a small garden such as this, and is formed from a few simple curves. The same shapes are used in both parts of the garden in order to give the design unity. The front garden plan is the only one that we have shown in the book – it needs to be simple and easy to look after. A mixture of shrubs and herbaceous plants in the borders provides for birds and butterflies; some larger shrubs and small trees form the backbone of a tiny 'woodland edge' continued at the side of the house. This is surfaced with bark chippings and accommodates both the repositioned garden shed, and the log-pile.

the females bear the berries. With wild holly (also an excellent insect tree), it is relatively simple to ensure that you have at least one male present, and you will probably have wild ones not far away. The cultivated hollies, however, often exist as either the male or the female form, but not both, so it is essential to obtain the female varieties for most of your planting. To complicate matters further, the varieties named 'something queen' are male, while those named 'something king' are female! A few varieties, such as *Ilex x altaclarensis* '*J.C. van tol*' or the golden-leaved form 'Golden van tol', are self-fertile so these present less of a problem and should produce berries regularly.

Other birds, especially some of the finches, prefer drier seeds rather than fleshy fruits. Some of these are included in our list, but the possibilities are endless. If you have a large garden, there are many suitable native trees that produce fruits and are attractive to birds. Birches (silver or downy) are excellent for redpolls, siskins and others; alders (preferably the native *Alnus glutinosa*) are fed on by the same small finches; pines are liked by crossbills and oaks by jays; beeches are a favourite with bramblings and chaffinches, and ashes attract bullfinches.

For the smaller garden, such beautiful plants as the wild teasel are excellent, especially for goldfinches, as are the thistles, though these are a little more problematical. The two most prolific and suitable thistles are spear thistle and creeping thistle, but they are apt to be invasive (especially the latter) and are legally classed as weeds in Britain. You could – if you are brave enough – collect thistle heads from elsewhere; take thick gloves and secateurs with you and hang the heads over a paved area. Some spare seeds may blow into the garden, but most should be eaten.

SUNFLOWERS
Sunflowers are very attractive annuals with their beautiful huge golden 'soup-plate' flowers. They are enjoyable to grow if you have children, who can measure their rapid growth rate, compete with each other over the height of different plants and watch them turn their flowers towards the sun. Given reasonable weather, they also produce a surprisingly large crop of fresh sunflower seeds in the autumn. However, these do not ripen until the foliage has turned grey-green and begun to die back, and they look rather odd, standing out far higher than anything else in the garden. The best plan is to cut the heads off as they ripen and hang them up somewhere convenient and visible in the hope that they will be found by the seed-eaters, especially greenfinches.

Nesting birds
Attracting birds to feed is relatively easy, but persuading them to nest in your garden is quite another matter. Birds bring food to their young until they can fend for themselves unlike insects which, in general, simply deposit their eggs on a suitable food source. So they need to nest in the middle of a territory which will keep both the adult pair and the voracious developing nestlings well supplied with suitable food. This inevitably means that your garden, which will form all or part of this territory, has to be reasonably varied and capable of providing at least some of the right food. We come back to the analogy used at the beginning of this book, of a jar of flowers in a concrete yard attracting a few butterflies which feed briefly but do not stay; similarly a nesting site, whether natural or artificial, in a sea of concrete is unlikely to be used by anything.

Most gardens, however, can form part of a territory of many common garden birds and the more varied and insect-rich your garden is, the more possibilities there will be for territories, and the more species-rich your garden is, the more birds will include it in their territory.

Common sunflowers Helianthus annuus, *growing in a garden.*

NEST-SITES

The next requirement for successful breeding is a suitable nest-site. Each species naturally has different requirements, and some are very fussy while others are more catholic in their tastes. The integrated approach advocated in this book will, in itself, provide a number of suitable nesting places for different birds, but you can also supplement whatever your garden has to offer with artificially created ones, mainly in the form of boxes. In areas where a lack of suitable nest-sites is the limiting factor, boxes are likely to be very successful. There are some excellent publications on the different types of nest-boxes; these are available from organisations such as the British Trust for Ornithology and the Royal Society for the Protection of Birds (see page 150 for addresses) and it is best to refer to these for more information. You need to decide which birds you are most likely to attract, based on your knowledge of the area and what you may already have. For example, if you have several resident families of tits breeding in holes in an old wall, it is unlikely that you will gain much by putting up tit-boxes. Also, some of the more specialist types of box may be for species that do not occur in your area. A good field guide to birds will tell you what you can expect locally and a county bird report will be better still.

Common Whitethroats are one of the more conspicuous warblers, often seen singing from a bush top, and they will come into gardens to breed if you provide suitable conditions in the form of a good thick hedge.

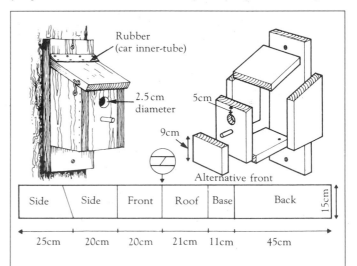

Side	Side	Front	Roof	Base	Back
25cm	20cm	20cm	21cm	11cm	45cm

A basic design of nestbox for hole-nesting birds, which can also be easily modified, by cutting away part of the front panel into a robin or flycatcher box. The size of the entrance hole is important if you wish to exclude house sparrows and other aggressive nesters, and it should be no more than about 2·75 cm. It can be reinforced with metal (not aluminium) if needed to exclude predators such as woodpeckers.

CATS

If you are trying to encourage birds and mammals into your garden, there is no doubt that cats can cause a problem, whether they are yours or your neighbour's. A recent survey in a Bedfordshire village showed that the 70 cats studied brought in 1090 animals during one year, consisting of 22 different species of birds and 15 species of mammals. These figures only relate to prey brought into the house, and it can be assumed that many other animals killed were never seen. It was estimated, for example, that the cats killed about a third of all the sparrows in the village in the year that they were studied. You can help the cats and the wildlife to co-exist better, with a little care. If the cat is your own, a bell round its neck will give warning of its presence, but this is rarely fully successful. Site feeding tables and nest-boxes so that the birds have a clear view of potential predators. If the cat is an unwelcome visitor from elsewhere, discourage it regularly – cats do not appreciate being drenched with water! As general rules, young cats are the most active predators; good weather is more likely to tempt them into the garden: and cats prefer to stay at home when it is raining.

You can encourage breeding birds by putting up bird-boxes, or adapting or improving any potential nest-sites you may already have. For example, if you have an outhouse or garage with exposed rafters, it will probably be suitable for swallows, but it will be of little use if they cannot gain access at all times. It is certainly worth leaving a side window permanently open or making an access hole above the door for these

BIRD-BOXES

The positioning of bird-boxes is critical if they are to be successful, and the most common reason for boxes being unoccupied, whether in gardens or elsewhere, is poor positioning. Generally, they should be placed in a shady place, either fully shaded or protected from hot sun and south-westerly wet winds. They should be reasonably well hidden but accessible by the prospective occupants. If possible, they should be sited at least 2 metres above the ground to avoid attack by ground-based predators, especially small boys!

beautiful birds. If there are no exposed beams, you could quite easily rig something up to suit them. They have lost so many of their farm nest-sites as old barns have given way to closed corrugated iron ones, that a little help from householders is very welcome. If you have a wall, try removing a brick or two, preferably somewhere sheltered by a bush or creeper, to allow robins or pied wagtails to nest. You can also try pruning your shrubs to provide more cover and strong forks for potential nest-sites. Best of all, plant a good thick mixed hedge of hawthorn, blackthorn, holly, beech and other shrubs and watch for results.

NEST MATERIAL

It is useful and enjoyable, but not essential, to provide some nesting material for birds. Sufficient material is normally available, but if you put out wool, feathers, dried grass and small sticks on a specially constructed table, or hang them up in a net you will be provided with some interesting watching, and save the birds time and effort. An area of mud in the garden, either on the edge of a pond or produced by 'puddling' a patch of the vegetable garden with a hose at the appropriate time, may attract house martins and swallows, as this can be in short supply with more farm tracks being metalled. These muddy places are in great demand in spring, and we have often seen 10 or more swallows or martins on one tiny patch of mud near houses.

One way of putting out nesting material for birds to use involves constructing a covered wooden table as shown below.

A group of brown long-eared bats roosting in the loft of a house. After the tiny pipistrelle bat, this species is probably the most frequent visitor to houses.

Mammals in the garden

There are relatively few mammals in Britain, or indeed in nearby continental Europe; and most of them tend to be nocturnal in their habits. As a result, they are little known and people often have very mixed feelings about them. Some, like hedgehogs, are almost universally welcomed, while others, such as brown rats, are equally widely hated. There are people who will gladly accept any mammal which comes into the garden, even if they cause damage, while others will wish to attract only certain species. The general ideas in the book will make a garden that is suitable for a wide range of smaller mammals, but it is also possible to make provisions specifically for some of the more popular or threatened species.

BATS

Bats are among the most threatened mammals in Europe, and they are also amongst the most misunderstood. Far from being dangerous, disease-ridden, damaging or smelly, they are attractive, small, furry, insect-eating flying mammals that need all the help that they can get. Over the past few decades, the changes in our countryside have greatly reduced the amount of food available to them, and the widespread use of insecticides has decreased their survival and fertility rate.

Perhaps worst of all has been the greatly increased use of toxic insecticides and fungicides in the treatment of roof timbers every time a house changes hands. Bats are highly sensitive to such chemicals, which they absorb directly or from licking their fur, and many have died or ceased to breed as a result. Over the centuries, bats have become increasingly associated with houses and other buildings, as their natural roosts in caves and hollow trees have disappeared or become unavailable; so it is inevitable that changes in the treatment of house roof spaces have hit them badly. Although bats are now strongly protected by law in Britain and there is a great deal of effort being put into finding alternative ways of treating timbers, there is still a considerable problem both from previous treatments and from legal or illegal loopholes, added to the fact that populations are already at a very low ebb.

Some homes will already have bats in the roof or under hanging tiles. If you want more information on bats or if you have a problem as a result of them, or want to do any work in the roof space that could affect them you should contact your local Nature Conserv-

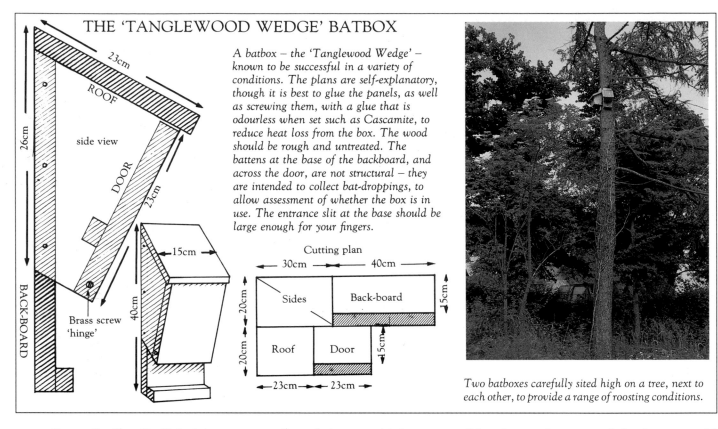

THE 'TANGLEWOOD WEDGE' BATBOX

A batbox – the 'Tanglewood Wedge' – known to be successful in a variety of conditions. The plans are self-explanatory, though it is best to glue the panels, as well as screwing them, with a glue that is odourless when set such as Cascamite, to reduce heat loss from the box. The wood should be rough and untreated. The battens at the base of the backboard, and across the door, are not structural – they are intended to collect bat-droppings, to allow assessment of whether the box is in use. The entrance slit at the base should be large enough for your fingers.

Two batboxes carefully sited high on a tree, next to each other, to provide a range of roosting conditions.

ancy Council office (in Britain) to arrange for advice, which will be given free.

BAT-BOXES

Bat-boxes can be a way both of helping the bats and adding interest to your garden. Although such boxes have been used in some countries for many years, particularly in forest areas, they are still a relatively new idea in others, and their value in gardens – as opposed to commercial conifer forests where no alternative roost-sites exist – is not fully known. If they are well designed and sited, especially in areas that lack roost-sites, you may be successful.

Bats need a number of different roosting places, including summer daytime roosts, winter hibernation ones and breeding sites, and they will probably have several of each – especially the first. Winter roosts are the most difficult to provide because of the need for considerable insulation, but better designed boxes will undoubtedly become available in due course. A design for the most successful type of bat-box is given, though you can also buy ready-made ones from a number of suppliers.

Once you have your boxes, you have to decide where to site them, and this is even more critical than for bird-boxes. As a general rule, it is best to position them as

high as possible above the ground both to avoid predators and because some species, such as the noctule bat, prefer roosts at least 5 metres above ground level. Other bats will use them at lower levels, but this has no particular advantages unless you have nowhere else to put them. Summer roost boxes should be sited facing from south-west to south-east, so that they receive sun for some of the day, preferably in the morning. They should be free of branches that may obstruct the flight paths, though they may be more successful if not placed anywhere too exposed. If you have a suitable large tree, it is well worth mounting three boxes on it, at the same level, facing north, south-east and south-west. Bats frequently move to find different conditions, so they will be able to fly between the boxes with little expenditure of energy, making the whole 'package' more attractive. You can also mount boxes on house walls. No preservatives should be used on the timber of the box, as this can be toxic to the bats. This obviously means that they will not last as long, but they are intended as a safe alternative to the treated roof timbers.

If the boxes are suitably designed and sited, bats should find them readily enough. To begin with you should inspect them regularly for any sign of use. Lift the lid very carefully, as there may be bats attached to it, and look inside. In the absence of any bats, you may find droppings, which are small, dry, dark brown to black and probably scattered over the floor of the boxes, rather than in piles like those of a roosting bird. Once you discover that you have bats, you will need a licence from the Nature Conservancy Council (in Britain) to continue to look at them, so you need to contact your local office immediately. If no bats appear to have visited your box within a year, try siting it somewhere else that may be more suitable. If you find that your boxes are attracting other residents, like roosting tree creepers or nesting hornets, you may prefer simply to put up another box and regard anything else that you have attracted as a bonus.

Some species of bat can also be attracted by nailing untreated wooden boards to a wall using battens to leave a space of 20 millimetres between the wood and the wall. These should not be placed on the north-facing wall. If you have a cellar with external access to it, try replacing the cover with a grille or making a suitable hole to allow bats in and out. This may even prove suitable as a winter roost.

The species of bats that you are most likely to attract are the pipistrelle, the brown long-eared bat and the serotine (except in the north), though four or five other species are possible visitors.

HEDGEHOGS

One of the most desirable and attractive of garden visitors is the hedgehog, though it is not in particular need of conservation. They are largely nocturnal, so you can live with a hedgehog family for years without knowing it, but if you put out a plate of bread and milk or a little tinned pet food you will usually be able to observe them. If the offering is being eaten, and you are sure that you are not fattening up the neighbour's cat, check the plate at night with a torch. Hedgehogs are exceptionally tame when feeding and you will certainly see them if they are there. Once you have discovered whether you have resident hedgehogs or not, there seems little point to us in feeding them regularly, unless you particularly want to watch them. If left to their own devices, they will keep a reasonable check on your slug population, being one of the few animals prepared to tolerate the slugs' mucus secretions; whereas if you habitually feed them they will inevitably take less natural food.

You can, however, encourage your hedgehogs to stay and rear their family by providing suitable places for them to lie up during the day, to hibernate or to breed. If you have a large garden with trees and banks, you will probably achieve this anyway by leaving piles

A stage in the construction of a simple hedgehog house, insulated with straw, under a woodpile. Hedgehogs will readily use such constructions if suitably sited.
Below *Hedgehogs are frequent and welcome visitors to the garden!*

of leaves in hollows, but in a smaller garden it is worth providing something specific. Hedgehogs will often nest under a garden shed if it is dry and sheltered, and there is easy access, but if you do not have a suitable place, you can easily build a 'hedgehog house'. The basic requirements are for somewhere dry and moderately well-insulated, and these can be achieved in a variety of ways. An old wooden box, for example, covered in polythene if it is not very waterproof, and

then buried under leaves or logs, will make an excellent site. You will need a hedgehog-sized access hole (about 10 to 12 centimetres square) or, better still, a short wooden tunnel to keep the interior dry, and you can fill the box with bedding material such as dry leaves, grass or straw. If you construct it so that you can open the lid without too much disturbance, you will be able to check on the occupancy every now and then, though this is not essential and in any case, should not be done in very cold or wet weather.

BADGERS

Badgers are highly unlikely to set up home in your garden, unless it is either large or very suitable. In fact they may not always be welcome if they do breed in the gardens. But, if they are in your area, you can encourage them in to visit so that you can watch them. Badgers have gradually moved into suburban gardens in recent decades, though much less so than foxes, and you still are more likely to see them if you live in the country-side. They have a very large feeding range, perhaps up

Long-tailed field mice are common visitors to gardens with the right habitats, especially where there is a combination of long grass and woodland edge.

to 100 square kilometres, so you may well be within range of a sett, though equally, unless you are quite close by, it may take some while before your food offerings are found.

The main natural food of badgers is earthworms, but they are reasonably omnivorous and opportunist, and will take many other foods if available. Household food scraps, especially if meat-based, will prove attract-ive, though badgers are said to be particularly fond of peanuts and raisins mixed with honey or molasses. The sweetener, apart from making the meal even more attractive, will also help the badgers to detect the food from further away, as they have an excellent sense of smell. Try putting out a plateful at a distance from the house at first then, if it is taken, gradually move the food closer until you can observe it from the house or an outbuilding. Do not forget that they will need to have reasonably easy access, as they are essentially ground-based animals. Badgers rarely emerge until after dark so you need to be prepared to stay up late. A torch dimmed by red cellophane or even brown paper is useful for observing your visitors without disturbing them. However, if you detect any evidence that your feeding is harming the badgers such as when they cross a nearby road, stop immediately.

MICE AND VOLES

For other mammals you are likely to be dependent more on chance. Natural supplies of food from trees and bushes, such as hazel nuts, acorns, beech-mast and crab-apples, will probably attract mammals, and, with hazel nuts in particular, you can work out what is visiting by looking at the way in which the nuts have been opened and comparing the hole with illustrations in a good field guide. It is now possible to obtain or make nest-boxes for the common dormouse, but these are not normal garden visitors, and you are unlikely to acquire these attractive creatures unless you are close to woodland and can provide the right overall conditions in your garden. The beautiful little harvest mouse is by no means confined to cornfields, and, if you have large suitable breeding sites, such as common land, nearby, you may be able to tempt them into your garden by leaving stands of tall grass, especially reed canary-grass *Phalaris arundinacea*, for them to nest in. Recent survey work has shown that they are commoner than previously thought, and they do occur occasionally in suitable gardens.

One further way in which you may attract and be able to observe certain mammals is described under 'Reptiles and amphibia'; it involves putting out suitable boards or pieces of corrugated iron (see page 87).

GARDEN FOXES

Foxes have always entered gardens at night, in more rural areas, in their search for food. Recently foxes have moved increasingly into towns and cities, and today they may be expected in gardens anywhere and they have even become less nocturnal in their habits. In fact, it has been shown that urban foxes are more successful than their rural counterparts, and they live at a higher density with more families per square kilometre. Gardens are regular breeding sites for more established foxes. Mating activity takes place in midwinter, with a peak in January, and this is when foxes are most likely to be heard in gardens. Subsequently, the vixen selects a suitable place as an 'earth' in which to rear her cubs; this may be under the garden shed, or in a hole in a bank. If the earth is close to – or even under – the house, then the period when the cubs are growing up in early summer is likely to be a noisy one. It can be an exciting experience to have foxes in the garden, but it may also give rise to the possibility of strong smells from the earth so, if you have the chance, encourage foxes to use the margins of the garden. Have a good look around your garden, checking unvisited areas behind sheds or fences, and looking under buildings to see if you have any resident foxes. They are so inconspicuous that you can easily have them visiting your garden without your knowing it.

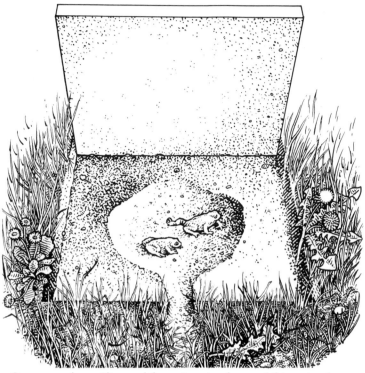

One way of providing a hibernation site for amphibians involves constructing a shallow depression, with access, under a paving slab, as described in the text.

Reptiles and amphibia in the garden

There are not many species of reptile or amphibia in Britain, or even in adjacent areas of mainland Europe. A few of them are very unlikely to venture into gardens, unless you live in a particularly suitable locality – for example residents of some villages among the Dorset heaths are fortunate enough to have sand lizards in their gardens – but most people do not.

The amphibians – frogs, toads and newts – all require water for breeding and tend to be associated with a wet environment for most of the year. Thus, the best way by far to attract them is to build a suitable pond, as described later. If, however, you are not intending to do this, you can still provide some help for amphibians, even though they will not stay to breed and it will be of little value if you are nowhere near a breeding site. As usual, a garden that is full of insect and other invertebrate life will tend to be most attractive to amphibians, as they are primarily insectivorous, and a good wildlife garden will have plenty of places for them to hide, hunt and hibernate. In more ordinary gardens, you can provide artificial sites that may be adopted by them for hibernation or other purposes. One good way

Slow-worms (above) are attractive and harmless lizards that frequently visit gardens where there are dry sunny banks and rough areas. You are most likely to find them under stones.

Common frogs (left) regularly use garden ponds in spring to spawn, though they will stay most of the year in a garden with plenty of food and cover.

Grass snakes are the most likely snakes to be found in gardens. They are harmless yet attractive creatures, as the picture shows, and readily distinguished from venomous snakes.

is by digging a shallow depression – 3 to 4 centimetres deep with a rising corridor leading up to ground level on one side; then cover this with a slab of stone leaving an entrance so that the corridor is just visible. You can lift the slab periodically to see what is there, but replace it carefully and do not look during very cold weather. Amphibians are most likely to use such sites if they are in damp grassland, but many other visitors may use them.

Reptiles – lizards and snakes – are generally more fickle, and less easy to attract since their breeding sites are more variable and not necessarily associated with water. The most likely visitors are slow worms (which are neither worms nor snakes but legless lizards) or grass snakes. Slow worms like dry banks with plenty of bare sunny areas as well as vegetated ones and abundant nooks and crannies. If you cannot provide this, a sheet of corrugated iron, or similar warmth-absorbing water-proof board, placed in a rough grassy area, will attract slow worms as well as mammals and inverte-brates. They are irresistible lures for many cold-blooded species, and surveys for the presence of rare

ATTRACTING GRASS SNAKES

Grass snakes are beautiful and harmless creatures, easily distinguished from the venomous adder by the lack of a zig-zag pattern on the back and by the conspicuous yellow throat collar. In June or July they lay their eggs, up to 40 at a time, in places where warmth is generated, such as manure heaps, rotting vegetation or, more relevantly, compost heaps the young emerge in August or September. The eggs are quite large, 25 to 30 millimetres long, and you may be lucky enough to attract a mated female if you provide a good compost heap that is undisturbed for the second half of the summer, as they will travel considerable distances in search of suitable sites. You may also be able to encourage common lizards by providing a bank with pre-made tunnels in it. (See page 126.)

snakes, like the smooth snake, are often carried out by putting down corrugated iron for them to hide be-neath. You will probably prefer to put the sheet down in an area well away from the house, and this will probably be best anyway particularly if you border any open country, rough land or a stream. If you periodi-cally lift the sheet, being careful to replace it gently in exactly the same place, you will discover all sorts of visitors including some of the smaller mammals that you had no idea were there.

CREATING NEW GARDEN HABITATS

There is a comment one often hears when discussing gardening for wildlife and the possible plants one might grow. It usually goes something like this: 'Plenty of those down the bottom of my garden – been trying to get rid of them for 20 years – I'll pay you to come and dig them up.'

Joking aside, this reflects the popular misconception that the wildlife garden is a garden out of control – allowed to run rampant. Although this is untrue, it does show that the techniques of wildlife gardening are not widely appreciated or understood. In fact, most of the disappearing countryside habitats, like hedgerow, coppice and hay meadow, were managed habitats. The impetus to 'borrow' them for the wildlife garden springs from the knowledge of what is happening to our countryside, and the possibility of re-creating some of these vanishing areas, in a setting which may be as small as a tiny urban garden.

Any form of cultivation interrupts the natural cycles of growth and decay and as a consequence the natural balance has to be maintained artificially. The wildlife gardener tries to use natural cycles as much as possible as a basis for management, and so some of these techniques may be slightly adapted or added to, but they are not difficult. Many of today's wildlife gardening techniques are taken from countryside management and from the continuing gardening traditions of the country estate. As in all gardening, it is a great help to know the reasons behind the methods, and in the case of wildlife gardening, this means understanding at least a little about how these natural habitats work. In Britain and Northern Europe, most of the habitats which can be recreated in the garden fall into three categories – woodlands, meadows and wetlands. The following chapters are about these habitats and making and managing them in the wildlife garden.

Opposite A newly-emerged skimmer dragonfly resting on pool-side rushes, with its empty larval case hanging below it.

Woodland and its edge

Woodlands are the richest and most diverse of all the terrestrial habitats in northern Europe. The number and variety of species within their three-dimensional depths is almost unbelievable, involving hundreds of flowering plants, at least as many lower plants, like lichens and fungi, thousands of species of insects and other invertebrates, and a high proportion of our native mammals and birds. Woodlands are also very old habitats. Sixteen thousand years ago, the vegetation of Northern Europe was just beginning to re-grow after the last major ice-age which had wiped the slate clean. As the climate warmed, waves of woodland began to colonise the tundra from the south northwards until, eventually, the whole area was covered with trees. Naturally all the mobile woodland species followed in the wake of this advance and, in a Europe which was still continuously linked by land bridges, most found suitable places to colonise.

Since this time, perhaps 10,000 years ago, the amount of woodland over most of the area has decreased, as clearance for grazing, hunting, cultivation and, later, for housing and development has proceeded apace, and now only about 10 per cent of Britain is wooded. Despite these continuing losses, there is still a huge flora and fauna which is adapted to this habitat, and any reasonably large and well-managed piece of woodland still supports an enormous number of species.

Nearly all the woodlands of lowland Britain and northern Europe have been managed for around the last 2,000 years, usually as coppice. Far from being the neglected wild forest of popular imagination, most medieval woods were very carefully and thoroughly managed to produce all the specific woodland products that the people of those times needed. Management by coppicing produces a particular and very interesting set of conditions. Coppicing is the practice of cutting a

A diagrammatic representation of the effects (left) of coppicing and pollarding on suitable trees or shrubs. The picture (left) shows a pollarded tree, cut regularly at about head height, then allowed to grow on – useful for keeping large trees manageable. In coppicing (right), the plant is cut right back to the base, to produce a broad bush.

Self-sown primroses, naturalised in a garden woodland glade, make an attractive display in spring, but also attract many early-flying insects.

shrub or tree back to ground level to allow it to produce a mass of shoots from its established root stock, thus giving a rapid production of small round wood. To maintain supplies, sections of a wood were cut in rotation; for example, where hazel was dominant, the wood was coppiced every seven to ten years over a series of compartments, so that some were always producing material for use. Thus the semi-natural woodland that covered much of the lowlands was constantly being cut, producing a series of glades or

edges, where a newly-created sunny patch abutted an older shady area. Flowers simply went dormant for a few years to erupt into a blaze of colour when the coppice was cut, allowing them to bloom in the absence of competition, while animals, such as insects, simply moved around the wood each year to the place where conditions suited them.

As a result, much of our native woodland wildlife is best adapted to an 'edge' or glade situation rather than to the depths of the wood itself, and this makes it much easier to recreate the rich woodland habitat within your own garden, without the need for an extensive area of trees. Paradoxically, the art of coppicing has virtually died out in Britain and it is now only practised in a few areas where demand (and the skills) still exist, or in nature reserves to maintain this important habitat. This means that many of our woodland species, especially those which lack the ability to become dormant like insects or birds, have become much rarer. The artificial garden woodland edge therefore offers a useful alternative to help counteract this continuing loss, though, of course, it can never replace real woodlands or offer a home to *all* woodland species.

As a habitat, woodland has certain peculiarities that its inhabitants particularly like, and it is worth mentioning these because the more you are able to recreate them, the more woodland species you will attract. It has a dappled mixture of sun and shade, giving suitable conditions for both sun- and shade-lovers; woods are humid sheltered places, protected from the extremes of heat or frost, from heavy rains and strong winds. They are also the last areas to dry out in a drought. There is a natural cycle of growth and decay, as the trees and shrubs reach deep into the soil for nutrients and then deposit them back onto the forest floor in the form of leaf or needle fall, or dead wood. Gradually the surface layers build up a mass of humus rich in nutrients, and this supports a huge interconnecting community of plants and animals feeding on rotting material or on each other. This is an important part of the woodland, and it is something you need to recreate, if you are to attract the full spectrum of woodland wildlife.

Woods have a marked structure, too, though this varies greatly according to management and woodland type. Most have some dominant trees which reach the canopy, such as oak, but they also have sub-canopy trees like field maple, cherry or crab-apple, while below these there is a shrub layer with guelder rose, hazel, holly, spindle and others according to circumstances. Many shrubs and small trees can form either the sub-canopy or shrub layer, depending on how well they are growing or how old they are. Below this, there are the herbs such as primrose, Solomon's-seal or bluebell, and beneath these, yet another layer with lichens, mosses, liverworts and others. The details vary and rarely occur in exact 'textbook' form, but the important point is that you can build up layers of woodland, and the more layers you have, the more birds, mammals and insects you will attract.

Lesser celandines naturalise very quickly in damp woodland in the garden, though they only flower profusely where it is not too deeply shaded.

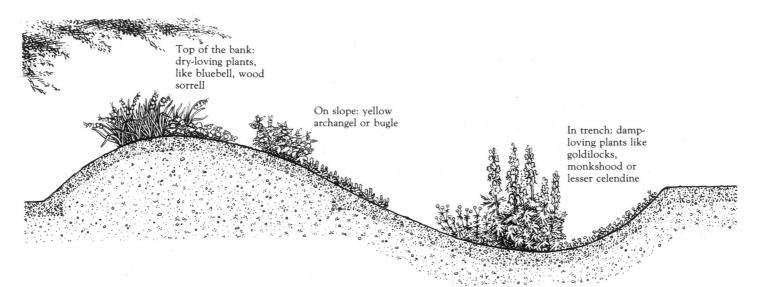

Top of the bank: dry-loving plants, like bluebell, wood sorrel

On slope: yellow archangel or bugle

In trench: damp-loving plants like goldilocks, monkshood or lesser celendine

An artificial feature of woodlands which is of particular historic significance and has become part of their ecology is the wood-bank. Wood-banks have served various functions over the centuries – to keep animals in or out, to mark ownership or help drainage. Whatever their origins, most ancient woods have them, and they often have their own particular ecology. The trees on the bank are usually very old, and frequently managed in a different way to those in the remainder of the wood; they may also be different species. The trench formed by the construction and upkeep of the bank is damper than the surroundings,

An artificial 'woodbank' allows you to use a wider range of woodland plants in the different conditions created. The species will vary locally.

Below, left: the dappled shade effect of a woodland glade or woodland edge is very attractive, and it provides ideal growing conditions.

and the slopes of the bank offer potential homes to woodmice, badgers, bumble bees, reptiles and many other woodland creatures. Though it will not have any historic significance (yet!), it is worth constructing some form of bank in your woodland edge to mimic this characteristic woodland feature.

So, a woodland edge is worth creating because it will be attractive, full of fascinating creatures from centipedes to garden warblers, and because managed woodland is a threatened habitat in the rest of the countryside. What is more, it is not too difficult to achieve even in a small garden.

Bringing woodland into the garden

The lower layers of a natural wood will change with the condition of the woodland itself, but also gaps in the over storey may occur, caused by one of the canopy trees falling. When this happens, the space – the woodland glade – is filled with temporary species until other trees grow to block out the light. It is this aspect that we can make most use of in the wildlife garden. You may be starting with a garden full of trees, often casting too much shade. The answer may be to take out just enough to open out a series of glades in the garden (check if there are any Tree Preservation Orders first). Alternatively, this type of effect can be created relatively quickly and easily. Woodland itself can vary in character – from open and dappled shade to dense and

A GARDEN IN A WOODLAND GLADE

This is one of the larger gardens we have featured, but individual ideas can easily be adapted to a smaller scale. The garden runs nearly E to W, the left-hand slopes being north-facing and shady, and ideal for a fern-covered bank with a woodland walk along the top. The main design is based on the idea of carving spaces out of a woodland, creating a series of warm, sunny glades, each with a different character, at right angles to the main garden axis. This garden has good views to the north, and while some shelter is necessary, the views are preserved by stretches of lower planting along this boundary. The glades are created by extending arms of planting from this side of the garden; one of these is continued into an area of rough grass with 'scrub', and the other by a wooden arbour and trellis.

A garden tour might take you up the steps from the patio and along the woodland walk to investigate the wildlife wood-pile under the trees. On the right, side-paths lead to a seat overlooking the garden. The main walk descends gently between the hedge and the coppice (where you can turn right down the log-steps) into the furthest garden glade, where fruit bushes under the trees provide berries for birds. A diversion can be made to the pond, before continuing back through the other open glades. Another seat, in the meadow-glade, has been placed where you can enjoy the meadow-flowers, as well as all the other garden sights and sounds.

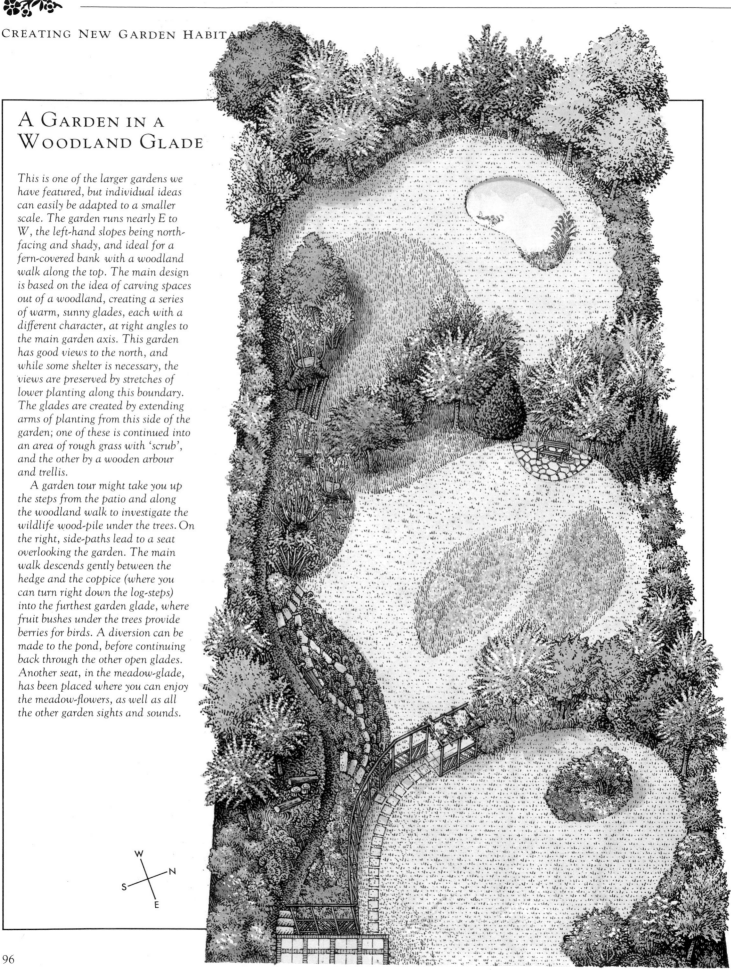

impenetrable – a mixture of both can be very good for wildlife and also very effective as part of the design. It can direct the view, or give a glimpse of another part of the garden through a partial screen.

TIPS FOR ESTABLISHING A WOODLAND

There are four key principles for establishing a good woodland and its edge.

● *Develop a good tiered structure that mimics the best of natural woodland structure, whatever the scale, to give scope for all the different colonisers.*

● *Build up a rich layer of rotting organic material, from simple leaf-litter to branches and logs or even stumps. This will develop naturally, but you can speed it up by bringing materials in (see below).*

● *Use as much native plant material as you can for the framework planting of the woodland, especially for the taller trees and shrubs. You can fill in later with the useful exotics, such as those producing nectar or berries, but as much as possible should be native.*

● *Make your basic planting as diverse as possible, using many different kinds of native trees and shrubs.*

In fact, the woodland garden can be a whole series of interconnecting glades which may have quite different characters. For example, you could have a pond in one, bearing in mind the rules for ponds and trees (see page 130) or perhaps a lightly shaded, ferny bank. Some woodland shrubs are more shade tolerant; others, of course, need more light. The difference in the light in the glade and round its edges allows you to intensify different kinds of planting areas. You can give even more contrast if you wish by accentuating the light and shade with corresponding foliage colours. Grass will grow in the clearings but not easily in deep shade, and this will grade into woodland ground-cover under the trees and shrubs. You can, however, use shade as part of the design, making use of areas of dappled shade under light-foliaged trees. What is often seen as a disadvantage can become a positive asset. Damp shade often supports lush growth which can be attractive.

The direction the edge faces is important. In larger woods, the southern edges are often the most species-rich. This is fine if your woodland edge has enough substance to it to give at least some shade at midday. If it

---

A damp shady woodland garden at Heale House, Wiltshire, which makes particular use of ferns for effect in a wild setting.

Comfortable, sheltered, and unobtrusive 'wildlife-watching' places are especially valuable at night. Alternative seats, even if only tree stumps, are useful if the wind changes direction.

is a very thin edge, you are better advised to angle it so that at least some of the site misses the hot midday to late afternoon sun, or you will be unable to grow the specialist woodland plants and your leaf-litter and wood piles will dry out and become useless.

However, for most people a complete woodland garden is unlikely to be the aim. Full canopy woodland is very shady, and it is difficult to grow many things successfully under mature trees, other than those that flower briefly in spring before the canopy closes over. A whole woodland garden also demands rather more space, especially as you are well-advised not to site it too near the house. Therefore, for most people, it makes more sense to recreate only the woodland edge. It can be fairly easily incorporated into a normal mixed garden as part of an integrated design, and will often relate to an existing boundary, perhaps softening the hard line by careful planting. If you have an existing large tree, then this can be used as the anchor for the

THE DELIGHT OF WOODLAND GLADES

One of the main attractions of natural woodland is that the view constantly changes as you walk through. You can emulate this in the garden by careful arrangement of the paths – gently curving – and the planting. As one glade opens into another, you might have a special group of plants highlighted against the shade, making it a lovely surprise to come upon round the corner. There might be a sweep of wood anemones, a stand of Solomon's-seal or a bank of dog violets in the spring. The dappled sun and shade at the edge of the clearing make a pleasant place to sit, but remember that if you need a hard surface in this area, it must be in keeping with the natural theme, as should paths and steps. Old stone, flat slices cut from tree trunks (watch out for slipperiness in wet weather) or wood chippings would all be appropriate. Gravel is good too – it can be rather bright to start with, but it gradually tones down and looks very natural.

woodland edge, but if you are limited by space, then you can still recreate this habitat by reducing everything in scale. Use birches, maples or rowans as your largest trees, instead of oaks or other large trees, but still try to retain the layered structure. There are very few gardens that cannot have some form of woodland edge, even though in some it may have to be greatly reduced in scale and extent. In a very small garden you can use trees or shrubs in a coppiced state, thus benefiting from having good native species without taking up a lot of space, or requiring a lot of work.

Planting and enhancing your woodland edge

Once you have decided on the positioning of your woodland edge, the next stage is the planting. If a completely new woodland is being established, it is very important to begin with clear ground (see page 49) to reduce competition around the new trees and shrubs until they are established. In most cases, this means some initial spot clearance, followed up by the use of mulch or black polythene on the immediate area around the trees. If you use uncomposted bark chippings as the mulch, both around the trees and generally, you are already beginning to build up the organic layer. Leaf mould is better in some ways, being more similar to real layers of the woodland floor but, until some shade cover develops, it will dry out in the sun, and it is better to incorporate this into the soil when planting,

and mulch on top with something more substantial.

Unless you are intending to build up a considerable size of woodland using major forest trees like oak and beech, it is best to plant your framework of trees and shrubs all at the same time and as soon as possible. Canopy trees tend to do best when growing through a layer of smaller trees, so this is a slower, longer-term business. The smaller trees are usually colonisers in their own right and they will grow readily without 'nurses'. If you do not have the resources to plant everything at once, it is probably best to get the trees in before the shrubs, because they have the most growing to do and will make the main framework.

SIZE AND SITING

As a general rule, smaller trees planted densely do better than a few larger trees; it is surprising how quickly they will establish, and they will often overtake larger trees which take longer to establish their root systems. If you are planting relatively few trees or large shrubs, it is best to go and select your own, rather than leaving the choice to the nurseryman, specifying healthy trees with a strong main shoot that has not been pinched out. This is not quite so important if you are planting a lot of trees. In either event, plant rather more trees than you need in the form of whips (small trees of about 1 metre high) then thin them out at a later date when they are established and you can see which are doing best. Tree shelters – the tall reinforced polythene

Wood anemones naturalise well in woodland gardens. This is the blue form of the native species, Anemone nemorosa robinsoniana.

· NATIVE TREES ·

	Soil requirements			Value and usage		
Damp	Well drained	Acid	Alkaline	Hedging	Coppice	Insects

Large trees only suitable for bigger gardens, or for coppicing where indicated

	Damp	Well drained	Acid	Alkaline	Hedging	Coppice	Insects
Ash	●	●		●		●	
Beech		●	●	●	●		
Wild cherry	●		●	●			
English elm		●	●	●	●		
Wych Elm	●			●		●	
Hornbeam	●	●	●	●	●	●	
Lime (small-leaved)	●	●	●	●		●	●
Oak	●		●	●		●	●
Black poplar	●	●		●			●
Scots pine		●	●	●			●
White willow	●		●			●	●

Medium-sized trees suitable for most gardens

	Damp	Well drained	Acid	Alkaline	Hedging	Coppice	Insects
Alder	●		●	●		●	●
Aspen	●		●	●		●	●
Downy birch	●		●		●	●	●
Silver birch		●	●		●	●	●
Field maple	●	●		●	●	●	
Yew		●	●	●	●		

Small trees sometimes treated as large shrubs

	Damp	Well drained	Acid	Alkaline	Hedging	Coppice	Insects
Bird cherry	●	●	●	●			
Crab apple		●	●	●	●	●	●
Hawthorn	●	●	●	●	●		●
Holly	●	●	●	●	●	●	
Juniper		●	●	●			
Wild pear	●	●	●	●			●
Cherry plum		●	●	●	●		
Rowan		●	●			●	
Wild service tree	●		●	●			
Whitebeam		●	●	●		●	●

tubes for protecting trees, also called 'Tuley tubes' – can be used; they speed up growth and cut down weeding.

Whips can be planted as close as 2 metres apart, or a little more if they are larger. Shrubs planted at about 1 metre apart will give dense cover in a surprisingly short space of time, especially if you encourage them to branch by hard pruning early on. A specimen shrub will need to spread to its full size and should have room to do so. Take care how close you plant trees to buildings. Some species are particularly notorious, especially willows, poplars, ash and elm, for their long invasive root systems which can eventually reach out to a distance greater than the height of the tree. It is a good rule to plant such trees at least one-and-a-half times their eventual height away from buildings; smaller trees, like rowan or whitebeam, are less invasive and need only be planted two-thirds of their eventual height away.

CHOOSING

Choice of species is obviously a subjective matter, and there is a reasonably wide selection despite the restrictions that your garden conditions will place on you. If you have existing trees, it obviously makes sense to try to incorporate them into your plan. Unfortunately, though, some species are unsuitable for creating species-rich habitats and they are better removed, preferably leaving the stump to rot *in situ*. Whether or not you do this obviously depends on various factors like the beauty of the tree, Tree Preservation Orders and its size. Sycamore, for example, is not a native of Britain or northern Europe and has very few resident insect species; it is also very invasive and produces a leaf litter which is not ideal for woodland flowers. Similarly, most conifers are rather unsatisfactory; though, if they are native conifers and you are in an area where they do well, you may benefit from trying to create a woodland edge that caters particularly for your local species. Rhododendrons are exceptionally popular in acid shady gardens; none are native to this part of the world though *Rhododendron ponticum* is widely naturalised. However, they are rather unsuitable as they tend to dominate the lower layers, allowing little else to grow. They are rather poor for birds and, apart from a brief glorious mass of colour and nectar, they are very dull for most of the year. The few deciduous ones are better than the evergreen species, but we know that bad cases of 'Rhododendron addiction' are hard to cure!

The tables left and on page 104 give a selection of tree and shrub species that could be planted, and they are grouped according to approximate final size to facili-

Rowan, or mountain ash, is one of the best small native trees, with attractive nectar-rich flowers followed by a large crop of bright red berries.

tate selection. Most species will grow reasonably well on any soil, if they are well looked after initially. If you have unusual soil, such as extremely acid or very wet, then you need to be especially careful.

Once the framework of native species has been established, you can add any non-native species that you particularly want, such as the ones which are good for birds (see page 75). A good general rule is to plant a wide variety of suitable species, removing any that fare badly later. Fruit bushes are excellent for a shaded garden, adding diversity; for example, blackcurrants, gooseberries and crab-apples (with 'John Downie' better than 'Golden Hornet') are all natural woodland shrubs, and you can even grow trees from pips, though you cannot predict their eventual size. Vines are excellent, either the wild type, or 'Brandt', with its coloured autumn leaves, are popular with birds. Raspberries are also very good for their flowers as well as fruit and they also tolerate shade. If you choose a short self-supporting variety, three canes planted together will soon form a clump which will soon spread, though it should not be difficult to control.

Guelder rose makes an excellent small shrub for shady places, with attractive heads of creamy flowers followed by clusters of berries.

SHRUBS

In the woodland garden, shrubs can represent two different things; either the understorey layer or, out on their own, the transitional phase known as scrub. You can use them in either way in the wildlife garden, and they will serve a valuable function in each. The scrub habitat, with a series of bushes in grassland, can be a very rich habitat for small creatures and for some birds – typical of the 'edge' habitats which are often richest in nature. Shrubs like guelder rose, alder buckthorn, elder and willows make good individual shrubs.

WOODLAND FLOWERS

The woody plants will make the framework to your garden and provide the obvious means of attracting wildlife, but it is the woodland flowers which provide much of the colour and life, as well as being prime attractants for insects. Plants have to be well adapted for growing under trees or even partially shaded by them, so you need to choose your species carefully basing the selection on those that do well in your local woods; the list on page 106 gives plenty of ideas. It is advisable to allow the trees and shrubs to begin to produce the more humid, shady conditions of an incipient woodland before you plant too many flowers.

A mass of planted and naturalised species flowering profusely in early spring under a woodland canopy that has not yet come into leaf.

Many will dry out in summer in the open, and they will almost certainly be smothered by light-demanding species. Unfortunately, almost all native woodland species, with a few notable exceptions like the foxglove, spread very slowly into new habitats, so they will almost certainly not find your developing woodland naturally. This means you have to buy them (NOT dig them from the wild) and plant them out yourself. Broadcast seed may not always survive and it is best to start some colonies with a few good-sized plants.

In the wild some plants, such as the familiar bluebell, tend to grow naturally in drifts, while others are components of mixtures. Plant the bulbous species, like wild daffodil, bluebells or ramsons, in bigger clumps to begin to establish your own 'drifts' of soft colour, but allow other species, like wood anemone, yellow archangel, wood sorrel or primrose, to intermingle and compete. The mosaic of woodland flowers in a natural wood is constantly changing, as subtle changes in light and soil give different species the advantage, and it is best to create this varied mixture in your own woodland edge. You can make special use of the early colour of many woodland plants which take advantage of the extra light available before the trees come into leaf. It will take time, and some species will fare badly while others may do too well, but, in due

course, you will end up with a mixture suited to your particular habitat – and it should look glorious every spring and early summer!

ADDING VARIETY
As time goes on, you can also build up the other aspects of your woodland garden – a pile of logs left to rot or the odd dead branch, and perhaps a hedgehog house in one corner. The wood is also quite a good place for the compost heap, as it will be well hidden and will not dry out in the shade. A woodland bank, as explained on page 95, is another good idea, and you can make the soil different in character from the remaining area. It will tend to be drier and slightly more acid anyway, but you can accentuate the difference by using different material. In due course, it will develop into a special feature of the woodland, colonised by different plants and maybe small creatures as well.

Managing the woodland edge
The management of woodland edges is not particularly difficult or time-consuming, but there is a little more to

do in the establishment phase. At first, you will need to look regularly at the way in which individual plants are spreading or failing, and they may need controlling, moving or boosting. Equally, you are likely to continue

· NATIVE SHRUBS · INCLUDING CLIMBERS

	Soil requirements				Value and usage			
	Damp	Well drained	Acid	Alkaline	Hedging	Coppice	Insects	Shade
Blackthorn	●	●	●	●	●		●	
Bog myrtle	●		●					
Bramble	●	●	●	●			●	●
Broom		●	●					
Alder buckthorn	●		●	●			●	
Purging buckthorn				●			●	
Sea buckthorn		●						
Bird cherry	●	●	●	●				●
Dogwood	●	●		●	●			
Elder	●	●		●	●			
Gorse		●	●					
Guelder rose	●			●	●			●
Hawthorn	●	●	●	●	●		●	
Hazel	●		●	●	●	●	●	●
Holly	●	●	●	●	●	●	●	●
Honeysuckle	●	●	●				●	●
Ivy	●	●		●			●	●
Mezereon	●			●				●
Wild privet	●	●		●	●		●	
Dog rose	●	●	●	●	●		●	
Field rose	●	●	●		●		●	●
Sweet briar	●			●	●		●	●
Spurge laurel		●	●	●				●
Traveller's joy	●	●		●			●	
Tutsan	●		●	●				●
Wayfaring tree	●		●			●	●	
Goat willow	●		●		●	●	●	●
Grey willow	●		●	●	●	●	●	●

to add plants as material becomes available or as gaps appear. Unless you are very short of space, the trees and shrubs can usually simply be allowed to spread naturally, with just a little cutting if they start to encroach on something else.

DAPPLED SHADE

In due course, however, you may begin to get a problem, depending how the plants have grown, as the edge hardens up and you have a sharp delineation between deep shade and full sun. To perpetuate the dappled shade effect that is so important, you will need to cut the lower layers back regularly. You can either mimic the effect of coppicing, and cut different shrubs right back to ground level every few years, or you can be selective and prune back branches of individual shrubs each year. Coppicing takes a little more nerve, but virtually every native tree and shrub (except the conifers, and wild cherry) coppices very successfully, and it prolongs their lives indefinitely if continued. Allow any individual shrub or tree to grow on for at least five years, preferably slightly more, before

This shows how you can establish your own hazel coppice in even a small garden; wild strawberry makes an attractive ground-cover that provides food for wildlife.

Natural ancient coppiced woodland in Suffolk, with oxlips in the foreground, and the recently coppiced and pollarded shrubs behind.

coppicing it again. As mentioned above, the technique is an excellent way of fitting in a wildlife-rich native plant which would otherwise be too big. The ancient technique of pollarding is also a useful one to remem-

PLANTS FOR WOODLAND AND · WOODLAND EDGE ·

If suitability is limited by soil type, preferences are indicated by the following letters: N neutral A acid C alkaline, calcareous M moist or wet	Full sun	Dappled shade	Full shade	Soil preference
Bluebell *Hyacinthus non-scripta*		●	●	A–N
Bugle *Ajuga reptans*	●	●	●	M
Butcher's broom *Ruscus aculeatus*		●	●	
Columbine *Aquilegia vulgaris*		●	●	N–C
Foxglove *Digitalis purpurea*	●	●	●	A–N
Ground ivy *Glechoma hederacea*	●	●	●	
Great woodrush *Luzula sylvatica*			●	A
Stinking iris *Iris foetidissima*		●	●	N–C
Lords-and-ladies *Arum maculatum*		●	●	
Lily-of-the-valley *Convallaria majalis*		●	●	
Monk's-hood *Aconitum napellus*		●	●	M
Primrose *Primula vulgaris*	●	●	●	
Solomon's-seal *Polygonatum multiflorum*		●	●	
Stinking hellebore *Helleborus foetidus*		●	●	N–C
Sweet violet *Viola odorata*		●	●	N–C
Water avens *Geum rivale*	●	●	●	M
Wild daffodil *Narcissus pseudonarcissus*	●	●	●	
Wood anemone *Anemone nemorosa*		●	●	
Wood spurge *Euphorbia amygdaloides*		●	●	
Wood avens *Geum urbanum*		●	●	
Wood sage *Teucrium scorodonia*		●	●	A–N
Wood millet *Melica uniflora*			●	
Wood vetch *Vicia sylvatica*	●	●	●	
Woodruff *Galium odoratum*		●	●	
Yellow archangel *Lamiastrum galeobdolon*		●	●	

ber; this involves cutting individual larger trees off at about head height from which point they will then regrow. Originally, it was used as a means of providing a supply of small wood in areas where grazing animals would have browsed any new shoots – a sort of head-height coppicing. In the garden, though, it can be a useful method of restricting the height of larger trees, while regaining their shade effect and value for wildlife. It is best to first pollard before the tree becomes too large, as it then becomes an expensive and difficult operation in the confined space of a garden. Once cut back, a tree should be re-pollarded every 10 to 20 years, depending on the space you have available, to prevent it from becoming too top heavy. Again, virtually all native trees will pollard successfully, except those mentioned as not for coppicing.

Brambles can become a problem in a wood or woodland edge spreading rapidly and killing, or at least hiding, all the flowers you are trying to encourage. It is not easy stuff to deal with, but in a small area you can

HONEY FUNGUS

Honey fungus is a real concern to gardeners, and there is the problem that bringing in wood-bark chippings and dead wood could increase the risk of its occurrence. If your garden is primarily one for wildlife, this is a problem you can just live with; but, if part of your garden is given over to specimen exotic shrubs collected on holiday in China 30 years ago, then you will be more concerned. Check any material that you bring into the garden for 'bootlaces' which will be reddish at the ends if actively growing. The main whitish body of the honey fungus occurs under the bark, so lift a piece carefully to check for signs – there is no need to remove all the bark; and logs from obviously healthy trees will not need examining. Burn any infected logs. Keep a close watch for the fungal fruiting bodies (use a good identification guide as there are many broadly similar species), and make every effort to contain it immediately if it does arise. It is generally reckoned that shrubs or trees growing strongly can resist its attacks, so make sure your woody plants are in a healthy state, with a good supply of nutrients.

A good hedgerow is a valuable feature at any time of year, and if you establish a variety of shrubs and climbers, you will find that it also looks very attractive. Here a hedge with black bryony berries, blackberries, ivy in flower, and other plants, is shown in autumn when it is a magnet for late insects on the nectar, and migrant birds on the berries.

keep on top of it by cutting it right back as soon as possible and digging out any larger roots; keep your eyes open for any new suckers or seedlings and remove them immediately. In shady parts, bramble has very little value for wildlife as it will not flower, but in sunny positions it can be one of the very best wildlife plants, producing marvellous nectar flowers, very edible fruits and foliage that is eaten by moths. The matter is further complicated by the fact that there are many strains or species of blackberry, and they vary in their value for wildlife and their invasive nature. Since they are not usually identified, or even identifiable, when you get them, it becomes largely a matter of trial and error.

Hedges

Hedges are a vital feature of the countryside, forming corridors between woods, marshes and other habitats among a sea of intensive arable land. They are also rather like thin woodland edges and may share some of the woodland edge species. Although they have traditionally served a particular purpose in the countryside – as windbreaks, stockproof barriers or boundary markers – there is no reason why they should not find a place in many gardens. They are one of the easiest and most controllable ways of producing something of the woodland environment, and they can also provide

shelter from the wind and mark the edge of your land.

Although never as rich as a good woodland, hedges can be remarkably good for wildlife. For example, some 20 species of butterfly may breed in hedges including some very rare ones. Many countryside birds, such as yellowhammers, linnets, whitethroats, garden warblers (and many more if trees are present) nest regularly in hedgerows, often at high densities. They are also homes or highways for many mammals and insects. Hundreds of species of plants can be found along hedgerows including both woodland and meadow species.

If you already have a hedge, you are lucky. You can

A HEDGE BOTTOM

This is the sort of hedge bottom that you should be aiming for in your garden! This particular example is a natural hedgerow in Dorset, many hundreds of years old, and the beautiful and intimate mixture of yellow archangel, bluebell, red campion, wild garlic and other plants has developed gradually perhaps from a former woodland flora, although obviously you cannot immediately achieve anything like the ancient gnarled layered ash trunk in the background, it is surprising how quickly a hedge will establish, and if you plant a suitable range of herbaceous plants (see list on opposite page) they will quickly form an attractive carpet that will form a striking feature in spring. These types of flowers will do best with some shade, rather than full sun, so they need to be established on the more northerly side of the hedge.

PLANTS FOR THE HEDGEROW EDGE
· AND FOR LONG GRASS AREAS ·

If suitability is limited by soil type, preferences are indicated by the following letters: N neutral C alkaline, calcareous D dry, well drained	Full sun	Dappled shade	Full shade	Shorter grass only	Soil preference
Betony *Stachys officinalis*	●	●	●		
Bush vetch *Vicia sepium*	●	●	●		
Dame's-violet *Hesperis matronalis*	●	●			
Dog violet *Viola riviniana*	●	●	●	●	
Everlasting pea *Lathyrus sylvestris*	●	●			D
Garlic mustard *Alliaria petiolata*	●	●	●		
Hedge woundwort *Stachys sylvatica*		●	●		
Herb-Robert *Geranium robertianum*	●	●			
Lungwort *Pulmonaria officinalis*		●	●	●	
Musk mallow *Malva moschata*	●				NCD
Nettle-leaved bellflower *Campanula trachelium*	●	●	●		N–C
Primrose *Primula vulgaris*	●	●	●	●	
Red campion *Silene dioica*	●	●	●		
Saw-wort *Serratula tinctoria*	●	●			
Greater stitchwort *Stellaria holostea*	●	●			
Lesser stitchwort *Stellaria graminea*	●	●			
Sweet Cicely *Myrrhis odorata*	●	●			
White dead-nettle *Lamium album*	●	●			
Wood crane's-bill *Geranium sylvaticum*	●	●	●		

either maintain it as it is, or use it as the basis of a richer woodland edge. To achieve this, check the species in the hedge and clearly mark any that can be allowed to grow on to become trees, making sure that you, and your neighbours, will be able to cope with them once they are mature. These will grow much more quickly than planted trees, because they can draw on an established root system. You can also plant small trees and shrubs in front of the hedge, after letting your side grow wild for a year to see what is there and how it will behave.

However, if you do not already have a hedge, you can soon establish one, with the advantage that its position can be planned. Aspect is critical with hedges, as with woodland edges, and you need to take this into account. An east-west hedge will have a very sunny southern side (more so than the woodland edge as there will be no overhanging trees to provide shade), while the north side will have more woodland character, with little sun. If only the southern side is in your garden, you will gain little but shelter from the north winds. The ideal aspect is probably a south-east/north-west alignment, with your garden on the north-east side, so that your hedge bottom is shaded from the hotter afternoon sun. A curving hedge will provide a good variety of conditions. Small hedges can be used as dividers within the main part of the garden; they will not be as good as proper wildlife hedges, but they will still be useful. Fruit bushes make good small hedges, if they are planted much more closely than for usual cropping requirements, and trimmed with a hedge clipper as necessary. The cheapest way to establish this type of hedge is to put in a row of hardwood cuttings. If you want them to provide a barrier, against cats for instance, use prickly bushes like gooseberries or worcesterberries, taking care not to plant them where children play regularly.

Mixed planting makes the most interesting hedge and, in the countryside, the oldest hedges have the most shrub species in them with 10 woody species not being uncommon. It is best to use quickthorn (another name for hawthorn) as a basis, with a generous mixture of other suitable species, such as blackthorn, spindle, field maple, privet and elm according to the soil. The quickthorn should be spaced evenly through the hedge, but the other species can be grouped if desired. It is easiest to dig out a trench when planting, then put in the plants as a double staggered row, each plant being 30 centimetres from the next, with the rows about 15 centimetres apart.

Fruit and nuttery!

For the less committed wildlife gardener, orchards and nutteries can be a reasonable compromise between fruit or nut production, visual appeal and wildlife value. They are rather like an open woodland with some of the layers missing, or a shady grassland area, depending on the density of trees. Orchards used to be interesting places for wildlife, with older fruit trees

The crab-apple called 'John Downie' is one of the best all-round wildlife plants, as well as being an attractive autumn-fruiting shrub.

providing homes for many nesting birds, including the extremely rare wryneck, and the pasture below often rich in flowers and studded with ant-hills. Nowadays, however, commercial orchards are very different with carefully managed trees, planted close together on dwarfing rootstocks; the grass is mown tightly between the rows, herbicide is used around the base of the trees, while regular doses of fungicide and insecticide are aimed at the fruit and foliage.

In your garden though, you can compromise between fruit production and wildlife value much more easily. It matters less if fruit production is lower, if there is some dead wood on the trees or a few blemishes on the fruit. You can manage the ground below satisfactorily as semi-shaded grassland, selecting species that like this type of situation, such as wild daffodil *Narcissus pseudonarcissus*, autumn crocus *Colchicum autumnale*, wood anemone *Anemone nemorosa*, bugle *Ajuga reptans*, wild tulip *Tulipa sylvestris* (not too shady, or it will not flower) and primrose *Primula vulgaris*.

If you are establishing a new orchard, you will, however, need to keep the competition around the base of the trees down for several years, preferably by heavy mulching. Because successful wildflower grasslands require low-nutrient soils, it makes sense to put organic manure around the bases of the trees to improve their nutrition without spoiling the remainder of the sward. You could also try mixing the traditional fruit trees with some that are particularly suitable for wildlife, such as wild crab-apple *Malus sylvestris*, or the best derivative, 'John Downie', wild pear *Pyrus communis* or *P. cordata* or common hazel *Corylus avellana*.

MANAGING A HEDGE

As the hedge establishes and matures, it will need to be managed. The way in which it is cut has a critical effect on its wildlife value. They are best trimmed once a year, preferably in winter, and certainly not during the birds' nesting season. If you have a double-sided hedge, you can cut each side at different times. The hedge should be thicker at the bottom than the top, so that it gradually assumes an 'A' shape, with plenty of cover at the base. As for size, a recent survey found that a hedge 1.4 metres high and 1.2 metres wide supported the most breeding birds. Do not tidy the bottoms up too much – leave plenty of leaf litter, seed-heads and other dead matter to attract hedgehogs, other small mammals, overwintering insects and visiting birds.

At the bottom of the hedge, you can plant in a few native herbaceous species that are naturally found there, such as garlic mustard, red campion, lords-and-ladies (wild arum) or greater stitchwort (see list on page 109).

It is doubtful whether snowdrops are really native plants except in a few areas, but they naturalise readily and provide an early source of nectar.

The flowery meadow

The origins of the meadow take us back to the last chapter – to the forests that once covered all of Europe – and the forest glades. The glade is a natural feature; a brief interlude; when for a few years, the sun reaches down to the forest floor. This chapter though, is about the *clearing* in the forest – or what began as a clearing – way back when early man first took his axe to the trees. The words 'glade' and 'clearing' appear to be interchangeable, but the clearing is *man-made*, and if the forest is not to flow greenly back again, needs to be *man-maintained*. These first clearings were gradually enlarged, maybe used for grazing, or for hay for winter feed; maybe hedged or walled about to keep stock from straying. All our grasslands have their beginnings here – these spaces, although originally artificial, have become an accepted part of the patchwork of our natural countryside. The modern ideas of creating garden meadows take us back full circle to those ancient clearings with their fleeting tapestry of flowers.

The appearance of the old-fashioned meadow was determined partly by where it was, and partly by the grazing or cutting regime that was carried out. These not only influenced which flowers and grasses grew but also their time of flowering. Re-creating a meadow in the wildlife garden therefore means rather more than leaving the mower in the shed and, once again, the more you know about the reasoning behind the recommendations, the more likely you are to succeed.

There are several types of meadow or grassland that may have a place in even a non-wildlife garden. We have already mentioned the flowery meadow, but many people also recall wayside flowers along country roads, and cornfields so full of flowers that one could hardly see the crop. We can add to these the rocky outcrops and drystone walls of upland grasslands and, although not exactly grass, the heather that covers so much of our heathland and moorland.

The great thing about flowery meadows is their incredible density of flowers, not just the number of flower heads but also the variety of species. In a long-established meadow in the countryside, you can find as many as 30 different species in a single square metre,

and in a whole meadow of 3 to 4 hectares, a total of 150 species of flowers is not uncommon. On the dry, low fertility soils of chalk downland where no single species can become dominant, as many as 40 or more species occur to the square metre. It is true that situations like these have built up over hundreds or even thousands of years, but it is still possible to recreate something of this incredible tapestry by careful planning and management. You can never recreate every aspect of an old grassland, for the soil will have slowly developed under the sward, ant-hills take centuries to grow to a particular size, individual plants in old meadows may be centuries old and the invertebrate fauna will have

---

A MEADOW OF FLOWERS

For most people the great attraction of having a wildflower meadow must lie in its extraordinary beauty. Whether you think of the mass of pastel-coloured flowers jostling together for space at the height of their flowering in June, the seed-heads clothed in dew in autumn or the sight of grass moths rising at dusk they are all worth having.

gradually developed – above and below ground – alongside the plants. But you can certainly have a go, and it is perfectly possible, using the experience accumulated in the last decade or two, to produce a superficial likeness of a meadow quite quickly.

There is another reason for attempting to recreate meadows. No other semi-natural habitat in lowland Britain, has disappeared so quickly. It is estimated that 95 per cent of our wildflower meadows have disappeared in the last 40 years and the other details of our countryside have changed beyond all recognition. We are not suggesting that tiny patches of scattered grassland in gardens can ever replace real ancient

An old hay meadow, or a well-established garden meadow, is a marvellous mixture of many different species, all jostling for space.

meadows and pastures, but they can undoubtedly help in many ways. They can act as staging posts for invertebrates, especially the insects, that need these areas; they are reservoirs of seed of declining species; and, the more people try to create wildflower meadows, the more we learn about the process.

Choosing the best site

There are a number of aspects to consider when planning where to put the meadow area in your garden.

POSITIONING YOUR MEADOW

Most meadow areas are better away from the house to give the prospect of looking out across a mown lawn to the rougher area beyond, perhaps with a hedge or trees beyond that. If the garden is large enough, you could reconstruct that feeling of coming unexpectedly on an almost forgotten, secret place, by siting it beyond a visual barrier and making it more like a woodland glade or old enclosed hayfield.

First, there are the purely practical considerations. Unless you are going to plant a mixture specially selected for shady areas, such as under orchard trees, the meadow should be sited in full sun as far as possible. In the wild, many of the most attractive meadows are often small and enclosed by high old hedges; these meadows are very evocative to recreate. The shelter they give is of great benefit to insects which love warm sunny places. While you obviously cannot plant high hedges all around your own meadow, you can at least site it where it will benefit from any existing shelter and perhaps add a hedge along one side. It follows, too, that it is better if the meadow area is quite small to retain the enclosed feel and to maintain the element of shelter.

From the design point of view, flowery grassland areas are best sited away from other flower areas, as neither would benefit visually. They are best set off against a stronger edge, such as a hedge or woodland edge – a flowery meadow running up against a shadowy hedge looks beautiful. From the ecological point of view, too, such a situation is ideal as many species use aspects of both open meadow and shady hedge or woodland; you cannot really have a complete 'woodland edge' effect without another habitat to compliment it. If you have open woodland with no shrub layer, more like an orchard, then you will need some visual definition between the two, such as a bark path or a strip of shorter grass. It is preferable, though, both visually and ecologically, to allow the grassland area to gradually grade into shrubs and on into taller trees, and this is not a difficult effect to maintain by occasionally cutting back the shrubs. You can also allow your flowery grassland to run up to a pond or wetland, and again they make excellent complimentary habitats, though it is best for practical reasons to have one side of the pond clearer for access.

You can maintain one part of your lawn as a flowery meadow with a different regime from the rest if you like, as the owners of this Dorset garden have done.

It helps to define the meadow area, by mowing an adjacent area tightly, as in this well-established wildlife garden at Winllan.

First steps: the flowery lawn

Perhaps the easiest way to start creating a flowery meadow *is* to leave the mower in the shed, at least for some of the time. Most people's lawns are full of a lot of other things besides grass, and they spend a fortune trying to get rid of them. Many of these are quite low plants – speedwell, daisies, self-heal, and cat's-ear – that would not actually persist in long grass. They will however, spread and flower in a fairly short sward. If you *do* only one thing – raise the height of cut of your mower; and *don't do* two others – use any chemicals (weedkillers or fertilisers) or mow more often than once a fortnight – this is a good start. Some species, like daisies and speedwells, will flower anyway under this regime, but it is better to allow a break in the mowing for four to five weeks in early summer, as this will allow some of the other flowers to bloom and set seed. Bird's-foot-trefoil will also grow in short grass but, if you are hoping to use it as the food-plant for common blue butterflies, the turf should be very short – under 2.5 centimetres. You may worry that neighbours will think you have been idling away your time in a deck-chair – the solution to this, and to edging the flower-beds, as

Short flowery lawns can look beautiful, and there are many flowers that can occur in them. Above left: a long-established turf with masses of spring sedge and bugle; above right: an easily-established mixture of slender speedwell and daisies; while the white clover, left, is ideal for bees – though bare feet should watch out!

well as to the possibility of any 'weeds' spreading into them, is to surround your flower lawn with an immaculately mown sward. If, however, the grass runs up to a hedge or informal shrub planting, you will not need to do this, and it might look strange if you did. Flower lawns like this are easy to establish and maintain. You can add other low-growing plants (pot-grown) if you wish, but generally the sward improves on its own as time goes by.

The flowery meadow

The feature of a wildlife garden that most often makes people want to go out and make one is the flowery meadow. This sight, with its romantic associations, immediately inspires one to rush out and buy corn-flower and poppy seed. It is at this point that you should sit down and think, because 'rushing' is defi-nitely not the way to go about it. The old meadows, so nostalgically remembered, full of kingcups and lady's smock, crane's-bill and goat's-beard, although van-ished it seems in the twinkling of the farmer's eye, had existed for hundreds of years. This complex commu-nity cannot be restored with a packet of seed alone. The extra ingredients for meadow gardening are threefold – low fertility, rigorous preparation, and time.

Fertility is important because, if it is high, grass will succeed at the expense of the flowers. Although a natural meadow community may occur on soils of varying fertility, these meadows have been unculti-vated for so long that the surface layers into which the flowers seed are relatively low in nutrients. The management regime in the first year is aimed at allowing the flowers to establish so that they can

· SUGGESTED GRASSES FOR MEADOWS ·

Brown bent *Agrostis canina*
Fine bent *Agrostis tenuis*
Crested dog's-tail *Cynosurus cristatus*
Downy oat-grass *Avenula pubescens*
Red fescue *Festuca rubra*
Sheep's fescue *Festuca ovina*
Meadow barley *Hordeum secalinum*
Meadow foxtail *Alopecurus pratensis*
Rough meadow-grass *Poa trivialis*
Smooth meadow-grass *Poa pratensis*
Quaking grass *Briza media*
Sweet vernal-grass *Anthoxanthum odoratum*
Wavy hair-grass *Deschampsia flexuosa*
Yellow oat-grass *Trisetum flavescens*

compete with the grass. This is also the reason why you should never sow rye grass (used for hard-wearing lawns) in a wildflower meadow – it is far too vigorous.

The *rigorous preparation* applies to the soil. You will almost certainly have problems establishing a flowery meadow unless you first get rid of as many weeds as possible. Docks and creeping thistle, if very abundant, can cause a continuing problem and are best avoided; but thistles will respond to systemic weedkiller treat-ment in the previous season. A fallow season before sowing is ideal, and the treatment recommended with black polythene (see page 49), although it takes a long time and looks ugly in progress, works very well.

The *time* refers not only to this preparatory inter-lude, but also to the following years. You will give the flowering plants the best chance of establishing them-selves if the meadow is cut regularly during the first year and you wait until the next for your flowers. As the meadow matures, you may want to add other species to broaden the plant community and, if these flourish, they will gradually spread out. It may, however, be several years before your meadow really looks the way that you had imagined.

When you start, you must always consider the fertility of the soil. It may already be low enough, such as on an unfertilised lawn on poor soil, but if you are

A summer meadow area, dominated by common spotted-orchids, cut once a year in autumn after they have set seed.

using an area that has been fertilised or used to be a vegetable garden, you will need to reduce the fertility. If you are working from an existing area of grass on dry, light soils, mowing and removing the cuttings over a season will probably be satisfactory. On damper or heavier soils, this will not work because of the deeper reservoir of nutrients. You can immediately reduce the fertility of a site by stripping off the turf and using it to make compost for application somewhere else. You can also strip the topsoil off, then either accept the lowered level or bring in subsoil to make up the ground again. Although you can use up some of the stripped topsoil elsewhere in the garden (maybe to make a bank in the woodland), the whole business of earth-moving can be a very daunting task over anything but the smallest of areas.

A better solution to the problem of soil fertility may be to plant for one year, or preferably a few seasons, an easy nitrogen-hungry crop that puts on a lot of growth. You can either grow this as a normal crop, or better still, broadcast the seed to give a total cover and then crop the resultant vegetation before it becomes too woody to make good compost. If you use something like mustard or cabbage, you can expect to take several crops off in a season by starting early in the year.

This may seem like a nuisance when you really want to get on with creating the meadow, but it cannot be stressed too highly that, in most circumstances, preparation and the *reduction* of soil fertility is essential to establishing a varied and stable meadow.

Sowing the seed

Before selecting the most suitable seed for your situation, make sure that you know your soil type (see Chapter 2) because there are different mixtures for a range of situations. You can make up your own mixture if you are not happy with those available commercially (and many *are* unsatisfactory), but it can be difficult to obtain all the species you require. If making up your own mixture, avoid rye-grass, cock's-foot, tall fescue *Festuca arundinacea* and Yorkshire fog.

Prepare the ground as for an ordinary lawn, making sure you get a fine tilth and a firm surface, rolling lightly or treading. Sowing is best done in autumn, between the beginning of September and mid-October, preferably earlier in this period to give more plants a chance to become established before the cold weather comes. Autumn is better than spring, since some species will germinate in the autumn, while others will only germinate after a cold period to break their dormancy. If you do sow in spring the period from the end of March to early May is best.

If the grass and flower seeds are packed separately, it is best to sow the grass first, rake it in thoroughly, and then sow the flowers *without* further raking. Lightly roll or tread the surface so that the seeds are in good contact with the soil but do not bury them if possible, as this reduces the germination rate.

Sow the seed thinly at the recommended rate — increasing the amounts will not improve establishment but will diminish the chances of the slow-establishing species. Usually you would sow at three or four parts

The even distribution of seed is best ensured by setting out a grid, then sowing the strips one way and then across them the other way.

grass seed to one part flower seeds, with the higher ratio of flowers (ie 3:1) being preferable but more expensive. You will require roughly 0.5 to 1 gramme of flower seed and 1.5 to 3 grammes of grass seed per square metre with the lower rates usually preferable.

An even coverage is important, especially for the grass seed. The flowers matter less, as some variation looks natural and it will happen anyway. Even coverage is best ensured by mixing the seed well, preferably keeping the flowers separate from the grass and adding in a carrier such as silver sand if desired. Now divide the plot into strips with string, such that each can be easily sown by one straight walk along it. Divide the seed into two parts, and then divide each half of the mixture into the number of strips; for example if you have measured out four strips, each half of the seed should be divided into four lots, so you will end up with eight portions. Sow a batch of seed into each strip, giving as even a coverage as possible, then re-divide the plot into the same number of strips but at right angles to the first. Sow these again in the same way. This method removes the 'edge effect' caused by your sowing seeds in the middle of each strip.

As an alternative to the more straightforward sowing procedure, you can use nurse crops with your seed mixture. Sometimes the use of Westerwolds annual rye-grass is recommended, but this is not really necessary for garden use and has to be handled carefully or the 'nurse' can swamp your mixture on anything but an infertile soil. Another method involves the use of a high proportion of annual wildflowers to produce a display of colour in the first year, but this involves altering the optimum first year cutting regime if you want to see them flower. They need to be left uncut for at least four to five weeks in June to July to ensure flowering, though you can carry on cutting after flowering. The seed crop is not important as very few will survive in the following year's closed sward.

If you are lucky enough to be able to acquire hay from an existing flowery meadow, (though this is a rarity nowadays), you can spread it out over the prepared seed-bed and leave it there; this will help to shade out the annuals while encouraging the grassland perennials that you want. This emulates the old farming technique used in hay meadows of spreading the sweepings from the barn in gateways or feeding areas where the sward had become broken up.

Managing your meadow

Management in the first summer is critical to ensure successful establishment, but after this you can become more flexible according to your requirements. The sward will need mowing six to eight weeks after sowing (or the same period after the start of the spring growing season if you sowed in the autumn) when it is over about 10 centimetres high. Before mowing, roll lightly, to ensure that the seedlings are well bedded into the ground, then cut at a height of between 5 and 10 centimetres using an Allen scythe or rotary mower. Used with care, heavy duty strimmers can also work reasonably, but lighter models tend to get indigestion. Mow again when the sward reaches the same height in another six to eight weeks (remove the cuttings each time). Mowing cures most weed problems, though you can hand-pull or spot-treat with Tumbleweed.

By the second summer, you can begin to adopt a more permanent regime, though you need a degree of flexibility at this stage to cope with varying growth rates; expect to mow once or twice during that year. By the third year, you should be able to instigate your preferred regime; if the grasses are still very dominant, keep the sward to under 10 centimetres again.

In subsequent years, the type of management will vary according to what you want from your meadow. The suggestions here are only guidelines and need not be adhered to very closely, as all will produce something of interest and value. The main choices are a spring-flowering meadow, a summer meadow, or one with butterflies in mind, though these are by no means mutually exclusive. You can decide right from the start by selecting a particular mixture of plants, or you can plant a very diverse mixture and see what does well before making your choice. The plants we list are likely to succeed in meadows under most conditions.

PRE-SOWING TECHNIQUES FOR · WILDFLOWER SEEDS ·

The seeds of some plants may be difficult to germinate for one reason or another, especially if the seed has been stored. Two pre-sowing techniques may be useful here.

· SCARIFICATION ·

Scarification is used for those seeds with a hard seed-coat, which impedes moisture access from the soil. Large seeds can be carefully nicked with a sharp knife but, as the majority of wildflower seeds are small, the same effect can be achieved by rubbing the seeds carefully between two sheets of sandpaper. Seeds which will benefit from this treatment include:

Salad burnet	*Melilot*	*Trefoils*
Clover	*Rock-rose*	*Vetches*
Crane's-bill	*Rose*	*Vetchlings*
Black medick	*Restharrow*	*Other legumes*

· STRATIFICATION ·

Some seeds need particular conditions in order to break down the germination inhibitors which are naturally present. Stratification methods can be used to provide some of these artificially if the seed has been stored or sown at a different time of year than would occur naturally. Many seeds need a period of winter cold before they will germinate, and the method of mixing the seeds with damp sand in a polythene bag, and leaving in the refrigerator (not freezer) for six to eight weeks, should improve germination. Then allow it to germinate somewhere warmer as it is (keeping damp) and the seedlings can be planted out straight from there. Many species benefit from this treatment, but the following are those for which it is particularly important:

Angelica	*Mignonette*
Clustered bellflower	*Oxlip*
Bluebell	*Wild parsnip*
Broom	*Rock-rose*
Purging buckthorn	*Sea holly*
Burnet saxifrage	*Sedges*
Wild carrot	*Spindle*
Cowslip	*Sweet Cicely*
Dogwood	*Viburnum spp.*
Hemp agrimony	*Violets*
Honeysuckle	*Weld*
Juniper	*Woodruff*

A spring-flowering grassland area, dominated by naturalised daffodils at Furzey garden in the New Forest.

Types of meadow

SPRING MEADOW

A spring meadow, with or without extra bulbs, should be left uncut until late June if the grass growth is lush, or rather later if it is not, and then mown down to 5 to 8 centimetres. Subsequently you can either keep the sward to between 8 to 10 centimetres by repeated mowing (simulating the traditional aftermath grazing that often took place following the taking of the hay crop) or you can allow it to regrow and give another modest display of flowers, and then cut back once in mid- to late-September. The last cut is important if you want an impressive display of spring flowers the following year. Incidentally, meadow saffron *Colchicum autumnale* can be treated as a spring meadow plant because its leaves will grow during the spring flowering period, but you will need to leave the meadow uncut during August and September to see it flower.

SUMMER MEADOW

A summer meadow can either be left uncut until a single mow in mid- to late-September, giving it a rake-over at the same time to remove dead grass; or you can keep the sward mown down to about 8 centimetres until June, especially if there is a lot of growth, and then cut once more at the end of the season. The second method helps to keep down invading coarse grasses, like tussock grass, which tend to grow strongly and spread in early summer.

For either of these meadows, it is important to remove the cuttings after each mowing, after leaving them to dry; this helps to reduce fertility. If your meadow is too fertile, you can include a higher than average proportion of the semi-parasitic plants that reduce the vigour of the grasses, particularly hay rattle *Rhinanthus spp.*, eyebright *Euphrasia spp.* and lousewort or red rattle *Pedicularis palustris* and *P. sylvatica* where the soil is damper. They are a common feature of natural hay meadows.

BUTTERFLY MEADOW

If it is your intention to provide a meadow where grassland butterflies will breed and feed, then you will need to adopt a rather different management regime. Unless you have a particularly large area or are in a very favourable position, you cannot expect more than a

A green-veined white butterfly feeding at its caterpillar foodplant, lady's smock, in a butterfly meadow, managed as described above.

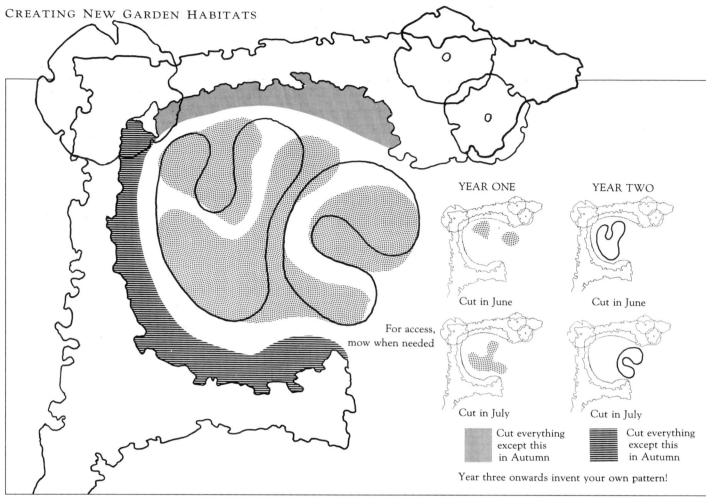

YEAR ONE

Cut in June

For access,
mow when needed

Cut in July

YEAR TWO

Cut in June

Cut in July

Cut everything
except this
in Autumn

Cut everything
except this
in Autumn

Year three onwards invent your own pattern!

proportion of the possible butterflies to colonise, but those that do will probably be successful, and make up in numbers what is lacking in variety. Each species likes different conditions, and you can best cater for a reasonable variety by making a mosaic of long, medium and short turf. Some butterflies also move from one height of grass to another as they develop. Very short turf, if it has bird's-foot-trefoil in it, will be suitable for the common blue and possibly the dingy skipper, in the right areas. For both, the grass has to be kept very short to allow the bird's-foot-trefoil to remain in a suitable state for larval feeding. Heavy mowing will not usually discourage the plant, even though you will cut off many flower heads, and the butterfly larvae feed mainly at night so they should be missed by the mower unless you mow by moonlight! It is better to use a Flymo type of mower to avoid too much rolling. These short grass areas should be kept fairly stable to allow the bird's-foot-trefoil and other short herbs, like thyme, to persist. The remainder of your selected area can be mown in a more random fashion to produce a mixture of long and medium grass, varying from year to year to prevent coarse grass becoming dominant. Establish some mown paths through the area each spring changing the route every year. Within the remaining blocks,

A suggested mowing scheme to provide a reasonable range of grass conditions in a small meadow area allowing several common species of butterfly to breed.

mow some parts in June and some in July, then give the whole area a final cut in the autumn, down to 5 to 10 centimetres and rake up the grass. Leave one part of the margin unmown for one winter and another part the next, to leave some standing seed-heads, especially if your meadow runs up to a hedge. This sort of regime, provided you have a reasonable range of grasses and are not overrun by rye-grass or tufted hair-grass, will provide suitable conditions for the more common grass-feeding butterflies, like the meadow brown, gatekeeper, speckled wood (if you have some shaded grassland) and perhaps the beautiful, but increasingly uncommon, wall brown. For further ideas on the subject, see the excellent book on butterfly gardening by Matthew Oates, listed in the Further Reading.

Adding extra flowers

Whatever type of meadow you are managing for, it is likely that over the years you will want to add to the species in the sward. This can work well and you will

MEADOW FLOWERS IN A GRASSLAND GARDEN

This sunny meadow-garden is on a dry, sloping site, but you can create one in many different situations. In this design, a large open space is sheltered by a thick country hedge which widens out occasionally to include other shrubs and trees of particular wildlife value. A certain amount of earth-moving has been employed to define two main areas within the garden.

A level lawn area near the house – the 'tamer' tidier part of the garden – includes a half-wheel herb border and an island butterfly bed, while the gradients of the 'wild' meadow below remain unaltered. The sloping bank which divides them is planted on either side with shrubs merging into the hedgerow, but in the centre of the garden, a rocky 'outcrop' and a dry-stone wall are bright with colourful rock plants. In the lower meadow, paths lead off from the 'flowery lawn' through the longer grass; one of them to a 'clearing' used for make-believe games by younger members of the family. The paths outline several areas of long grass managed by different cutting regimes, and a rougher outer margin where hedgerow-edge plants are allowed to dominate.

If you have acid soil, a heather 'lawn' like this one at Fiddler's Cottage, Hampshire can be created with persistence by means of an annual autumn mow.

OTHER FLOWERY HABITATS

There are other flowery 'natural' habitats that you can attempt to recreate, according to your garden conditions or particular interests. For example, if you are on particularly acid soil, especially if you know that it supported heathland before becoming a garden, it is an interesting idea to recreate a heathland 'lawn', like the one shown on the left. You can either aim for a pure heather lawn, using native heathers (such as ling Calluna vulgaris *or bell heather* Erica cinerea) *or you can mix in other heathland plants, like bilberry* Vaccinium myrtillus *or petty whin* Genista anglica. *A single annual mow, in autumn after the majority of plants have flowered seems to work best, but you do need infertile soil.*

If you live very close to the sea, and especially if you are on reclaimed coastal land, you can try mixing in some coastal species, such as sea-holly, sea bindweed, strawberry clover or hare's-foot clover, to make a maritime turf.

gradually move towards the best mixture for your soil type, though most meadow owners find that their range of flower species varies remarkably from year to year. You can put in additional species in several ways. Either grow your own turf by sowing the desired seeds in small seed boxes without grass, and then transplant these miniature turves into the grassland when ready. They will look odd at first, but will soon establish as a self-thinned clump, and begin to spread if they like the meadow. Alternatively, you can plant out individual pot-grown plants, bought or home-grown, into the grassland, preferably in autumn, clearing a little space for each as you do so. You can make drifts of species or aim at a mosaic, but either way you will need to select species which will flourish in a closed sward, using the lists of recommended species or observing what does well in meadows locally. Choose native species, rather than their cultivars if possible, or species known to naturalise well. You can add bulbs in the same way when the meadow is established, and these can be native species or not, according to what you are trying to create.

These methods can also be tried on an ordinary lawn if you do not want to go to the bother of preparation and reseeding, and are sometimes very successful. However, you will stand a far greater chance of success with a lawn on poor soil.

ANNUAL FLOWERS

Although annual flowers, such as poppies or cornflowers, are often mixed into meadow mixtures, they are quite unsuited to survival in a closed sward, and they

Former bonfire sites make excellent seed-beds for annuals, like poppies and cornflowers, in an otherwise closed turf, as shown in the photograph, taken at Winllan garden.

· RECOMMENDED WILDFLOWERS FOR LAWNS AND MEADOWS ·

If suitability is limited by soil type, preferences are indicated by the following letters:
N neutral
C alkaline, calcareous
M moist or wet
D dry, well drained

	Short flowery lawn	Spring meadow	Summer meadow	Soil preference	Easily-grown species
Agrimony *Agrimonia eupatoria*			●		
Wild basil *Clinopodium vulgare*			●		C
Lady's bedstraw *Galium verum*			●	●	DNC
Bird's-foot-trefoil *Lotus corniculatus*	●	●	●		
Greater bird's-foot-trefoil *L. uliginosa*			●		M
Black medick *Medicago lupulina*	●				
Bugle *Ajuga reptans*		●		●	
Greater burnet *Sanguisorba officinalis*		●			
Salad burnet *Sanguisorba minor*		●	●		N–C
Bulbous buttercup *Ranunculus bulbosus*		●			N–C
Meadow buttercup *Ranunculus acris*		●	●	●	
Cat's-ear *Hypochoeris radicata*		●	●	●	
Lesser celandine *Ranunculus ficaria*	●				M
Red clover *Trifolium pratense*		●	●		
Cowslip *Primula veris*		●			N–C
Meadow crane's-bill *Geranium pratense*			●		N–C
Cuckooflower *Cardamine pratensis*		●		●	M
Daisy *Bellis perennis*	●				
Ox-eye daisy *Chrysanthemum leucanthemum*		●	●	●	
Dandelion *Taraxacum officinale*	●	●			
Common knapweed *Centaurea nigra*			●	●	
Greater knapweed *Centaurea scabiosa*			●	●	C
Goat's-beard *Tragopogon pratensis*		●	●		
Autumn hawkbit *Leontodon autumnale*			●		
Rough hawkbit *Leontodon hispidus*			●	●	N–C
Mouse-ear hawkweed *Pilosella vulgaris*	●				D
Marsh-marigold *Caltha palustris*		●			M
Meadowsweet *Filipendula ulmaria*			●		M
Hoary plantain *Plantago media*	●	●			N–C
Ribwort plantain *Plantago lanceolata*			●		

	Short flowery lawn	Spring meadow	Summer meadow	Soil preference	Easily-grown species
Ragged-Robin *Lychnis flos-cuculi*		●			M
Common restharrow *Ononis repens*			●		D–C
Meadow saxifrage *Saxifraga granulata*		●			
Field scabious *Knautia arvensis*			●	●	
Small scabious *Scabiosa columbaria*			●		N–C
Self-heal *Prunella vulgaris*	●		●	●	N–C
Common sorrel *Rumex acetosa*		●			
Germander speedwell *Veronica chamaedrys*		●			
Slender speedwell *Veronica filiformis*	●				
Perforate St John's-wort *Hypericum perforatum*			●	●	
Thyme *Thymus drucei*	●				D
Tormentil *Potentilla erecta*	●				
Lesser trefoil *Trifolium dubium*		●			
Horseshoe vetch *Hippocrepis comosa*	●				C–D
Kidney vetch *Anthyllis vulneraria*	●	●	●		C
Tufted vetch *Vicia cracca*			●	●	
Meadow vetchling *Lathyrus pratensis*			●	●	
Yarrow *Achillea millefolia*			●	●	
Yellow-rattle *Rhinanthus minor*		●	●	●	
Bluebell *Hyacinthoides non-scripta*		●			
Wild daffodil *Narcissus pseudo-narcissus*	●	●		●	M
Fritillary *Fritillaria meleagris*		●		●	M
Meadow saffron *Colchicum autumnale*		●			
Snowdrop *Galanthus nivalis*	●	●			M
Star-of-Bethlehem *Ornithogalum umbellatum*		●			

have probably put more people off meadow gardening than anything else by their dismal failure in the second year! However, many highly attractive native flowers fall into this category, and they can be grown success-fully but not as a meadow. The essence of growing these annuals, which are most often weeds of cornfields, is annual disturbance so that a new seed-bed is produced each year. They are best grown in a small plot set aside especially for the purpose. Prepare and sow the same as for meadows (see page 118) but, at the end of the year, you need either to encourage them to self-seed by raking the site and shaking out the seeds or, better still, collect the seed and recultivate the site. One wildlife gardener we know has found that his bonfire sites produce excellent crops of annuals if a handful of seed is thrown on after it has cooled! Old-fashioned corn-field weeds make an interesting mixture to grow as an annual crop and, if you sow them with a suitable arable crop, they can look most attractive.

Rockeries and banks

Rock gardens and dry-stone walls provide a useful way of bringing in some material of a different chemistry into your garden, if you so wish, and they may also provide a home for other wildlife. A traditional

'rockery' would be out of place here – this sort of rock garden might resemble a flowery outcrop on a rocky hillside. If you build anything like this, it is a good idea to insert some short lengths of piping at the start, then pull them out when you have finished, leaving some holes for lizards and other animals. Planting will depend upon the type of rock you are using, but it can include a high proportion of native species and some suggestions are made in the table opposite.

The meadow is also the ideal site for a simple pile of loose rocks or stones. It needs to be piled up so that there are plenty of nooks and crannies available for future occupants.

Finally, it can be attractive to establish one or more flowery banks in the garden. You can build these up yourself, using either indigenous or imported material, and they can either be planted with the appropriate meadow mix if not too steep or uneven to mow, or you could establish a 'herb bank' using a dense planting of herbs such as hyssop, marjoram, thyme, lemon balm, basil and others, which are allowed to flower freely rather than being maintained for culinary use.

A rocky outcrop, with suitable plants growing wild among fine grasses, needs less maintenance than a traditional rockery.

Pinks and other plants established on a dry stony heap made to resemble a natural rock outcrop in a garden in Germany (above). Culinary herbs like Rosemary (below) are very attractive to insects if they are left to grow on and flower, and they make a beautiful aromatic feature as a bank.

PLANTS FOR ROCKY OUTCROPS · AND WALLS ·

If suitability is limited by soil type, preferences are indicated by the following letters: **N** neutral **A** acid **C** alkaline, calcareous	Walls	Rock outcrops	Soil preference
Bladder campion *Silene vulgaris*		●	N–C
Bloody crane's-bill *Geranium sanguineum*		●	N–C
Biting stonecrop *Sedum acre*	●	●	
Common fumitory *Fumaria officinalis*	●	●	
Harebell *Campanula rotundifolia*		●	
Herb-Robert *Geranium robertianum*	●	●	
Horseshoe vetch *Hippocrepis comosa*		●	C
Ivy-leaved toadflax *Cymbalaria muralis*	●		
Maiden pink *Dianthus deltoides*	●	●	
Red valerian *Centranthus ruber*	●	●	
Rock cinquefoil *Potentilla rupestris*		●	C
Sheep's-bit *Jasione montana*		●	A–C
Toadflax *Linaria vulgaris*	●	●	
Wild strawberry *Fragaria vesca*	●	●	

Water and marsh

Of all the small-scale habitats that you could create in the garden, the pond is probably the most effective at bringing in extra wildlife. Even a small one will add a whole new dimension to a garden, giving a home to a completely new range of plants and animals, as well as providing a drinking, washing and feeding area for many already present. If you do nothing else in the way of wildlife gardening, try making a pond on the lines suggested and you will be rewarded with dragonflies, damselflies, frogs, newts and birds coming to drink, and much more besides.

A good pond teems with life. On the surface there may be little to see except the leaves of aquatic plants and a few pond skaters, but the closer you look, the more you realise what is there, and pond-dipping is a fascinating activity for children or adults. If you take just a drop of water from a pond in summer and look at it under the microscope, you will see masses of very small creatures – both plant and animal – moving about, and these are the basis of a whole chain of life that exists under the water. A pond is a complex three-dimensional habitat, full of scavengers, primary producers, predators, parasites, herbivores, decomposers and many more.

Some animals spend their whole life in ponds – such as water fleas, water snails, and flatworms, while others are highly adapted to aquatic life and yet are also able to fly to new waters. These include many of the water beetles and the bugs like water boatmen, pond-skaters and back-swimmers. Then there are those which spend much of their adult life out of or away from water, but are tied to water by the need to breed there. Dragonflies, for example, spend all their larval life (often several years) under water but, when they emerge to become winged adults, they feed wholly out of the water, often well away from it. Amphibians are similar, spending much of their life in moist environments

Toads (far left) are regular visitors to well-established garden ponds, if conditions are to their liking, where they will spawn in spring, producing strings of eggs that are quite unlike frogspawn. One of the many groups of insects that will take up residence in your pond is the water boatmen, such as this lesser water boatman (left).

which are often, but not necessarily, near water; but they always return each spring to mate and lay eggs in the water which give rise to the aquatic young stages – the tadpoles.

Other animals have little to do with water most of the time, but they come regularly to drink, wash or find food. Many garden birds will visit a pond to drink, eat the aquatic flower seeds, pick off an emerging damselfly or just bathe and enjoy themselves; while you will probably be surprised at how many insects drink or use the water in one way or another. You may like to try arranging a hose-pipe a little way above the shallow part of your pond, so that the water splashes

Ponds that are designed with wildlife in mind can still look stylish, as this pond in Dusseldorf with its terracotta poolside planter shows.

into it, drop by drop. It is quite fiddly to arrange, and does not look especially elegant, but if you can get it right, you may be surprised at the numbers of birds that it attracts.

Over most of the lowlands, ponds are mainly artificial features. Although many were created by the receding ice after the last ice age and persisted because of the cooler climate, the natural tendency of ponds is to become colonised by plants and ultimately return to dry land in our lowland climate. Thus, virtually all of the open, still water that you see is either artificially created or maintained or, more occasionally, of recent natural origin. Ponds have been made to provide water for stock, irrigation for crops, fire-fighting, drinking and industrial water and various other uses, as well as being created entirely for amenity use in some places such as parks and gardens. So the idea of making ponds is nothing new. But, at the same time, the use and upkeep of small ponds in the countryside has declined greatly, as pumped water has become available and vast reservoirs replace the tiny individual supplies. Ponds, as a semi-natural habitat in the countryside, are therefore as threatened as most, and creating them in the garden environment can help to maintain populations of, at least, the more common aquatic animals, and provide links in the chain of aquatic habitats through the countryside.

Where to put your pond and wetland

Where you put your pond or wetland is very important, for all sorts of reasons, both practical and ecological; and it pays to think clearly about this before embarking on any work.

It is important to site any pond away from trees and,

EVEN SMALL PONDS WILL DO

Ponds are an immensely valuable habitat to have in any garden, and they can be as large or small as the situation demands – we have seen ponds of some 75 centimetres long by 35 centimetres wide that were full of plant and insect life. Ponds are useful on their own, but can be made even more attractive by careful siting in relation to other habitats, and in combination with a wetland fringe.

more especially, to ensure that there are no trees to the south and west. There are several sound reasons for this. Firstly, trees shed massive quantities of leaves, and the great majority that fall onto the pond will sink to the bottom. This layer of rotting leaves, if left, will spell death for the pond because it uses up all the oxygen in the water during the rotting process and smothers any plant life that might help to reverse the effect. You will eventually be left with a very dull pond that contains nothing but breeding mosquitoes and bloodworms – not at all what you set out to provide! At the same time, if the trees are to the south and west, they will cast a shade over the pond, and over you when you are down on your hands and knees looking in – ponds undoubtedly fare better if they are warm and sunny. Also, as the prevailing wind blows from the south-west in many

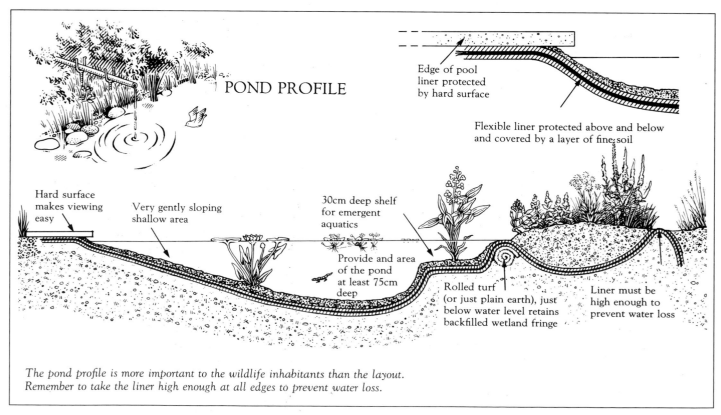

POND PROFILE

Edge of pool liner protected by hard surface

Flexible liner protected above and below and covered by a layer of fine soil

Hard surface makes viewing easy

Very gently sloping shallow area

30cm deep shelf for emergent aquatics

Provide and area of the pond at least 75cm deep

Rolled turf (or just plain earth), just below water level retains backfilled wetland fringe

Liner must be high enough to prevent water loss

The pond profile is more important to the wildlife inhabitants than the layout. Remember to take the liner high enough at all edges to prevent water loss.

Although ponds are best sited in a natural hollow, this is sometimes difficult, and some will need to be retained artificially. It is possible to make a feature of this, as shown here, with a built-up viewing platform and step.

areas, the leaves from any trees on these sides will be blown in, too. If your pond has to be near trees, make sure it is at least to the south of them, and preferably as far away as possible.

Whether the pond is close to the house or not is not usually critical, but there are a number of factors to be considered. If sited at a distance, it can make a nice feature at 'the bottom of the garden' to 'go and look at the pond'. It also makes it easier to isolate the pond from young children or dogs, if necessary. Ponds can be very dangerous for young children, who can drown in even shallow water, and this obviously has to be taken into account. Dogs will not drown, but they are very good at puncturing linings and ruining plants, so you may find it better to have the pond somewhere where it can be fenced off.

Where ponds occur naturally, they are likely to be in a natural hollow, and it looks best if you can site them like this in the garden. They look odd if the ground falls away steeply from them, unless part of a chain of pools with flowing water, and it is better if there is at least some level or gently sloping ground around. Any existing natural unevenness can be used and exaggerated. If the ground does fall away steeply, you can build up the lower side and site the pond a little away from it, disguising the effect with planting; or you might accept

the artificiality and make a feature of the drop in level by putting in a kneeling-level viewing platform.

If you are fortunate enough to have a stream in your garden which you can dam, you can create a pond without using a liner. This will greatly increase the amount of aquatic life that you already have in the stream, as many more species prefer still water. However, producing a pond will increase evaporation and change the flow rate, so you may need to check with neighbours and the water authority if the effect is likely to be significant. What would happen if the dam collapsed, for instance? You need to know the approximate annual flow pattern of your stream, and particularly whether it floods, dries up or produces vast amounts of silt at any time of year. Some streams are winterbournes (these are usually on chalk or limestone) which dry out totally in summer, and you may not be aware of this if you moved into a property in winter. Silt-traps can be installed as a means of reducing silt build-up in the pond from flowing water, as this can be a considerable and continuing problem. It is best to install the silt-trap at the same time as you build the pond, and you will have to check and maintain it periodically. For the pond, you will need some sort of sluice system to control the water level and possibly an extra overflow. If you wish to construct an artificial stream system, we suggest you take specialist advice as DIY results are not always successful. If you engage professional contractors, obtain a few estimates beforehand – these sort of operations are more expensive than most people realise.

The vegetation or habitats next to the pond are important, too. Obviously, if you are planning a whole wildlife garden from scratch, you can work these factors in from the start. But, if putting a pond into an existing situation that you do not want to change, you need to think carefully about the existing surroundings. Ecologically, it makes sense to have some wetland, such as a well-vegetated damp area, next to the pond, and a good deal of interchange between the two habitats will take place. Wetland animals fare better with a pond nearby and pond creatures fare better with a wetland nearby. It is also attractive to have some tall fringing vegetation on one side of the pond to look across at and perhaps to mask the edges. Species like dragonflies need upright plants to climb out on in their final larval stage. It is usually better not to have the wetland totally surrounding the pond, as you will want to gain access to the pond from at least one side, though you can construct a raised path if required. The trouble

with having wetland next to the pond is that the vegetation will constantly invade the water, so annual maintenance is required. Some hard surfacing will be needed at one side of the pond; although not ecologically valuable, it makes access and study easier, and protects the vulnerable pond edges. This could be paving or timber decking and, if the latter runs out over deep water, will need a safety handrail.

For some situations, a raised or partly-raised pond may be preferable. Although there are no reasons why one should not do this, there are a few provisos. You can and should include a small wetland area (you can work this into your design) contouring the base under the liner so that it shallows towards this edge. Raised ponds are inevitably more formal in appearance, because straight lines are usually involved, so are best placed nearer the house where they can fit in well with a patio, rather than in the wild garden area. They have the advantage of being easy to look into and, although they will not be quite as good for wildlife as a more natural pond, they can still be a valuable asset.

Constructing the pond

Having decided roughly where to put the pond, or ponds, you need to decide on the shape, size and construction method.

The size will be roughly determined by your available space, but it is worth remembering that a bigger pond will be richer in species and easier to maintain in a

---

PONDS BY THE HOUSE

If the pond is close to the house, it can be easier to watch the goings-on, such as birds bathing, without causing them disturbance. It is usually also easier to run water to the pond when filling or topping up. Even if using rainwater, and there are definite advantages in doing so (see page 140), it is easier near the house because you can make use of the larger surface area of your house-roof to increase the rainwater collecting capacity of your garden pool. You will need a water-butt with an overflow into the drain and a tap that will take a hose leading to the pond. As you need the water to flow into the pond, the butt will have to be raised up somewhat. A summer shower may not be enough to raise the water level to its normal height, but this strange arrangement works well enough if you are prepared to dash out and connect up the hose when it rains, and turn the tap off again when the pond fills up. The water-butt acts as a reduction valve between the drain-pipe and the hose, rather than a permanent reservoir. Another reason for siting the pond near the house is that it is simpler to arrange a power source if you wish to have water pumped by electricity. Remember that any features of this kind should be in keeping with the natural style of the pond.

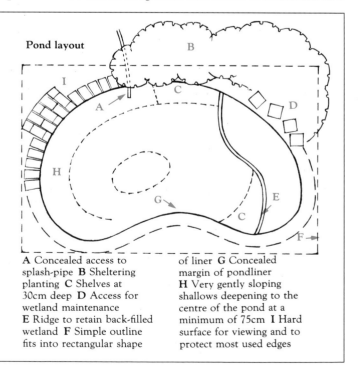

Pond layout

A Concealed access to splash-pipe B Sheltering planting C Shelves at 30cm deep D Access for wetland maintenance E Ridge to retain back-filled wetland F Simple outline fits into rectangular shape of liner G Concealed margin of pondliner H Very gently sloping shallows deepening to the centre of the pond at a minimum of 75cm I Hard surface for viewing and to protect most used edges

An artificial stream course in a garden, planted with predominantly native plants in a natural style, making an effective and attractive feature. Right: Edgings such as this can be very effective visually, but wildlife-access areas with gently sloping banks and adjoining vegetation, should also be provided.

balanced state; around 4 to 5 square metres is a good size and not too difficult to work with. Depth is important, too. The minimum preferable depth at some point should be about 0.75 metres, to give a variety of conditions and provide a section that does not freeze right through in winter. There should also be shallower areas as described later.

Shape is not critical, though it should fit in with the type of garden. An informal shape with easy curves will be best in an informal natural setting, but if your wildlife garden is a town courtyard, it may be very different. The pondlife will not mind the shape as long as you follow the other recommendations.

There are a number of different ways of constructing a pond, and which one you choose will depend upon your circumstances. The commonest method nowadays uses a liner of some type, though other methods are also discussed.

Puddled clay is rather limited in application, but very useful if you already have clay on the site. You

MAKING A GARDEN WILDLIFE POND

The stages in creating the simple pond shown here are similar for any pool, whatever lining is used. **Left:** *basic shaping gives a small wetland area to backfill later, and a variety of depths in the main pond – the deeper parts grade very gently into extensive shallow areas – check now that sides are roughly level.* **Centre:** *after clearing the bottom of stones, carpet underlay is a cheap alternative to tougher purpose-made protection – overlap strips well.* **Right:** *black polythene is a cheap temporary liner so no top protective layer is added, though it would normally be better to do so – the water surface provides a check on final levels.*

need a thick layer of clay, about 150 to 200 millimetres thick (usually put in as bricks), which has to be well trodden by many pairs of bare feet or boots first. When the water is added, the clay swells and produces a waterproof layer. A common problem with clay linings, however, is that if they dry out, they are liable to crack and cease to be watertight. Once this has happened, it can be difficult both to locate the leaks and repair them. Consequently, the pond needs to be kept topped up to prevent this from happening.

Bentonite is a particularly pure form of clay that you can buy in a dried form, ready to spread into a carefully graded pool basin. When you run the water in, (gently, to avoid moving the layers around), the bentonite swells even more markedly than ordinary clay and it makes an excellent lining. It is a good way of sealing large lakes, though it is best to obtain specialist advice if embarking on such schemes; there are companies who specialise in this work.

Concrete ponds need not be as bad as they are said to be, if carefully designed and planted. However, you need to pay critical attention to the shape of the bowl and margins, and also to the thickness of the lining in order to avoid winter damage. As they are best completed all in one go, you have to be prepared to work quickly. If you do want to use concrete, then we advise that you study the subject well beforehand. Leaks are very difficult to cure so you must get it right first time round.

Liners, of one sort or another, are probably the best

The overlapping edge of this butyl liner is disguised by a crazy-paving surround. This pond needs netting in autumn to keep out falling leaves.

solution for the average garden pond. We do not recommend the pre-moulded types, readily available at garden centres – they are usually the wrong colour, the sides are too steep for a wildlife pond, and too slippery to allow a build-up of silt, and they are difficult to integrate with a wetland garden. Also the larger ones, in the sizes you would need, are difficult to install well, and expensive.

This leaves the options of polythene, PVC or butyl rubber which are all flexible liners for placing in a pre-dug hole to mould to its shape.

The cheapest is polythene, but it is fragile and short-lived. If you buy 1000 gauge black polythene and double it over, it makes a reasonable base for a cheap pond. It does, however, tear and puncture easily, and breaks down and cracks where exposed to sunlight, so it is not really worth considering unless you cannot afford the more expensive alternatives or do not need the pond to last long.

PVC is commonly sold in garden centres, usually in blue or stone colours, though you can obtain black if you persist. It is more expensive than polythene, but will last rather better with an estimated life of five to 15

years. It is easily punctured, and, though there is a more expensive version which is reinforced with nylon, this is more likely to resist tearing than punctures.

The best option is undoubtedly butyl rubber if you can afford it. This comes in two grades of strength, and is reckoned to last 20 to 50 years, depending on the type. It is easier to handle than the others because it is less fragile, though it is also heavier. All three materials are, however, vulnerable to damage to some extent, especially when laying, so care needs to be taken whatever the choice.

The next stage is to mark and dig out the pond. The shape can be marked out using a hosepipe, but bear in

A recently-created pond, planted with native species like flowering rush and figwort, that is already looking well-established.

A WETLAND HABITAT GARDEN

A damp site is sometimes regarded as a drawback, but the wetland garden uses it to advantage. Many of these features could be created on a drier site, and on a smaller scale. The general effect here, is of a gently sloping damp meadow, sheltered by an enclosing hedge of wetland shrubs. The base-plan design for this garden is derived from the shapes made by a sinuous line curving back upon itself. The inter-relation of the resultant shapes reflects that of the complementary habitats within the garden.

There is a small grassy areas with spring and summer bulbs in the drier, upper part of the garden. This drier part is better for general use, and the paved and gravelled patio area is dry underfoot even in wet weather. Good paths are important in a wetland garden, not only to ensure dry feet, but also to protect the delicate wetland habitats. The main path round the garden is surfaced with hoggin and gravel, which will eventually blend into the surroundings. Moveable duck-boarding crosses the wetter areas. The wetter lower part of the garden includes a large damp meadow and extensive wetland planting. A short stream links the two pools, and a timber viewing platform allows the pond-life to be studied from both sides of the main pond.

This bold edge-planting is visually effective, as well as important for pond wildlife – water plantain and water flote-grass are emergent aquatics with bold leaf forms.

mind at this point the possibility of extending the liner to create a wetland fringe at one side of the water. It is best to avoid any tight curves or corners, as it will be difficult to fit the liner into them. When working out the depth and profile and the required amount of liner, there are a number of important factors to remember.

● Allow for the thickness of any paved edge (about 6 to 7 centimetres) or turf so that it does not finish up slightly raised.

● Allow extra depth for at least 3 centimetres of cushioning sand below the liner and about 2 centimetres of underlay, if using this; then about another 5 centimetres for the soil which will go on top of the liner as base substrate.

● Make the profile such that you have at least one area that slopes gently from deeper water to the edge, and include at least one area of shelf about 30 centimetres wide and 30 centimetres below the final water level.

To calculate the amount of liner to buy, you will need to know the width, length and maximum depth. For PVC and butyl, which stretch slightly, you will need length times width with twice the maximum depth

added onto each dimension, with no allowance necessary for the overlap round the edges. With polythene do the same but add an allowance for the overlap, since it does not stretch and is very cheap anyway.

When digging out it is best to strip the turf, if there is any, and keep some for edging the pond. Then excavate the rest of the hole leaving the shelves and shallows. Try not to disturb the soil below your base level, as it leaves a firmer bottom, but you can use a rake to get the shape right and remove projecting stones. Look carefully for any more sharp stones and remove these, then spread 2 to 3 centimetres of fine sand all over the floor as a soft base for the liner. Also check carefully all round the pond to make sure that the sides are level.

Polyester matting, sold especially for the purpose, can be placed under the liner though, if you can only afford one layer of it, it is best placed on top where it offers some protection from light and puncture damage, and makes a good non-slippery surface for soil to adhere to. It will need to be weighed down initially, as it tends to float to the surface. As an alternative, you can use old carpet underlay above or below the liner; or folded sheets of newspaper below as additional cushioning.

The next step is best done with two people. Unfold the liner over the pond and position it centrally. Hold the edges down gently with bricks or stones for the moment. If using butyl, you *can* walk on it, but it is best to avoid doing so unless it has protection on top; PVC or polythene should never be walked on. A small pile of rocks (or something similar) placed at the deepest part will weigh the lining material down, and can be left as a home for amphibians or other pond creatures. Alternatively, a few buckets of soil (riddled to remove sharp stones) can be used. You can now begin to fill the pond with water, easing off the weighted down edges as the pond fills up. As the water nears the top, check the level of the sides again and correct any discrepancy. Cut off any extra liner and underlay, leaving about 30 centimetres of both to be finished with turf or slabs.

You can now add more soil to the pool, to hold the liner and matting down, and produce a rooting substrate. The matting should prevent the soil from sliding down the sides into the deepest part. The whole thing will look a muddy mess at this stage, but it will clear surprisingly quickly. Leave it for at least a week before embarking on any planting.

You can then plant into the basal layer of soil that you have provided – they root very easily in standing water – or you can put some on the ledge areas in perforated pots from which they will root outwards.

Leave some of the shallow area and the ledges free of plants to allow access for animals. Different water plants require different depths of water, and ultimately you should aim for a good mixture of plants – those with erect stems and leaves, some that *may* be rooted which have floating leaves, and submerged 'oxygenating' plants. The floating leaves help to shade parts of the water and reduce algal build-up, in addition to providing cover for aquatic organisms, and mating, courting or viewing platforms for a whole range of insects. If you choose from the widespread and easy native plants such as white water-lily *Nymphaea alba*, frogbit *Hydrocharis morsus-ranae*, water-plantain *Alisma plantago-aquatica*, arrowheads *Sagittaria sagittifolia*, flowering-rush *butomus umbellatus*, yellow iris *Iris pseudacorus*, water-starwort *Callitriche spp.* and even duckweed *Lemna spp.*, then you will establish an excellent framework of plants. If you can persuade a friend to give you a few plants plus some 'sludge' from their pond, you should acquire a good range of adult,

Left: a dramatic way to both hide the pool liner and provide access across the pond – on a smaller scale, reduce the size of the individual elements. Right: these cleverly interlinking raised pools in the BTCV wildlife garden at Chelsea (designer – Tony Hitchcott) show that you can provide for pond life at any level.

An example of a highly successful pond created primarily for wildlife in a Shropshire garden, at Limeburners, using a butyl liner.

larval and egg stages of aquatic invertebrates to get things going. It is amazing how quickly it all settles down and how soon additional animals and plants begin to colonise the pond.

Pond management

Ponds need relatively little management, but there are a few tasks to do and points to watch out for.

You will almost certainly get algal blooms – masses of strongly growing green or bluish algae – at first. These thrive on high nutrients, and it is inevitable that a new pond will have water with high nutrient levels. These are partially caused by the high nutrients levels in the tap water used to fill the pond. As a general rule, however, you want to keep nutrient levels as low as possible to begin with; so use rainwater instead of tap water to top up the pond, and keep or clear leaves out

of the pond. But there are various other ways in which you can reduce the nutrient levels.

The problem tends to be worse in wide shallow pools where the water receives more light and warmth than in a deeper pool. By giving the pond a reasonable minimum depth, you will reduce the problem to some extent from the start. If you also remove any of the filamentous algae when they become too luxuriant, you will be removing some of the nutrients locked up in them at the same time. However, there will be large quantities of pond creatures in the weed, so you need to shake it out well, then place it in a pile by the edge or in a very shallow area. After a while, you can remove the weed completely (taking the top layer first if it is a high pile), and compost it for use elsewhere. You can also crop your emergent aquatic plants in the autumn to remove more nutrients, though you should never

· PLANTS FOR WATER AND WETLAND ·

	Oxygenating plants	Floating-leaved plants	Emergent pond plants	Plants for wetland		Oxygenating plants	Floating-leaved plants	Emergent pond plants	Plants for wetland
Angelica *Angelica sylvestris*			●		Meadowsweet *Filipendula ulmaria*				●
Amphibious bistort *Polygonum amphibium*		●	●		Monkeyflower *Mimulus guttatus*				●
Arrowhead *Sagittaria sagittifolia*			●		Pondweeds *Potamogeton natans, P. crispus*		●		
Bogbean *Menyanthes trifoliata*			●	●	Purple-loosestrife *Lythrum salicaria*				●
Brooklime *Veronica beccabunga*				●	Ragged-Robin *Lychnis flos-cuculi*				●
Cuckooflower *Cardamine pratense*				●	Royal fern *Osmunda regalis*				●
Duckweed *Lemna sp.*		●			Sedge – Bowles Golden *Carex stricta*				●
Flowering rush *Butomus umbellatus*			●		Spiked water-milfoil *Myriophyllum spicatum*	●			
Fringed water-lily *Nymphoides peltata*		●			Water avens *Geum rivale*				●
Floating water-plaintain *Luronium natans*		●			Water-crowfoot *Ranunculus peltatus and others*	●	●		
Frogbit *Hydrocharis morsus-ranae*		●			Water forget-me-not *Myosotis scorpioides*				●
Gipyswort *Lycopus europaeus*				●	Water mint *Mentha aquatica*				●
Greater spearwort *Ranunculus lingua*			●	●	Water starwort *Callitriche stagnalis*	●	●		
Great hairy willowherb *Epilobium hirsutum*				●	Water violet *Hottonia palustris*	●		●	
Hemp agrimony *Eupatorium cannabinum*				●	White water-lily *Nymphaea alba*		●		
Hornwort *Ceratophyllum demersum*	●				Yellow flag *Iris pseudacorus*			●	●
Marsh marigold *Caltha palustris*				●	Yellow loosestrife *Lysimachia vulgaris*				●

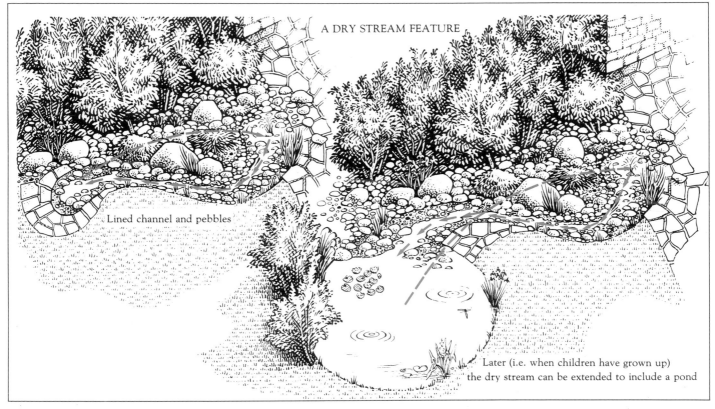

A DRY STREAM FEATURE

Lined channel and pebbles

Later (i.e. when children have grown up)
the dry stream can be extended to include a pond

A self-contained water feature like this, with a re-circulating pump in a sunken tank, can expand to include a 'real' pond at a later date. Take care that any joins or overlaps do not allow water to leak out.

remove more than a third, as the dead seed heads and tall stems provide overwintering sites for invertebrates, emergence areas for spring insects and food for birds and mammals. Algal blooms can also be reduced by ensuring a good surface coverage with aquatic leaves; this reduces the amount of light reaching the water and so helps to cut down algal growth. Avoid using an algicide if you possibly can, as they tend to upset the natural regime. If you periodically have to clear out the pond because of excess leaf litter, silt or invasion by plants like greater spearwort, then the same practice should be followed. Leave the vegetation by the pond or in shallow water until it has dried out, before removing for compost.

Although ponds are an obvious target for fish introduction, it is better to avoid bringing in exotic fish into a pond intended for wildlife. Those with sizeable fish populations rarely produce much in the way of wildlife, as the tadpoles and aquatic organisms are eaten all too quickly. A few small native fishes will become part of the system, but exotic fish, especially if you feed them, are liable to get out of balance with the rest of the wildlife.

Wetlands

'Wetlands' are areas of vegetation that are more or less permanently wet, but are not open water. They share some of the same plants and animals with ponds but have many of their own; and the combination of water and wetland in the garden (as in the wild) is a very strong one. There are many wetland plants, in particular, that do not really succeed well in open water, but thrive and look beautiful in a wetland area. These include marsh-marigolds *Caltha palustris*, purple-loosestrife *Lythrum salicaria*, yellow loosestrife *Lysimachia vulgaris*, water mint *Mentha aquatica*, common valerian *Valeriana officinalis* or marsh valerian *V. dioica* and many other attractive plants.

You need a reasonably large area of ground to make a useful wetland, though if it is too large it will make access difficult. If making the wetland at the same time as the pond and using the same liner, you need to incorporate it into your excavation and retain a low ridge between the two areas. The ridge top should be just below water level, to allow water to lap into the wetland, but high enough to prevent it from becoming part of the pond. The wetland plants evaporate water (transpire) from their leaves very rapidly in warm weather, so where the pond and the wetland are part of the same system, you will find that the plants tend to

A shady wetland garden can be successful, if you choose your plant material very carefully, but don't expect much life in the pond.

draw the level of the pond water down more rapidly, just as grass penetrating from the edge does.

A wetland can also be created away from a pond in a similar way. Dig a suitable hollow where desired (and you need worry less about the danger to children as there is no open water) and line it with polythene or PVC; polythene is satisfactory here, since no sunlight will reach through the vegetation and soil to degrade it. Push a few drain holes in the deepest part to prevent the wetland from becoming sour and anaerobic. You can trickle irrigate, but in most climates the impeded drainage caused by the liner will produce an area that is wet enough. You could associate this wetland with a dry 'stream bed' of water-worn pebbles and a few rocks just for effect. In fact, if you get the levels right, you could line the course of the stream, and use a bubble fountain to splash over the pebbles and down what would then be a totally safe watercourse. The water available would only be enough for a short distance but, later on, you could add in a pond at the lower end (perhaps when your children are older) to make a

complete water feature, though remember not to join the wetland to the pond, or you will lose all your water because of the holes in the wetland liner! Wetlands created like this, using a liner, will be more acid than your soil, so choose your plants accordingly.

Although the monkey musks are of American origin, they naturalise readily in northern Europe, and make an attractive addition to poolside planting.

PLANTS FOR YOUR WETLAND

Once you have created a wetland area, you can plant it up with a variety of plants, including any of the ones mentioned, together with bogbean Menyanthes trifoliata, *marsh cinquefoil* Potentilla palustris, *ragged-Robin* Lychnis floscuculi, *grass-of-Parnassus* Parnassia palustris, *yellow flag* Iris pseudacorus, *hemp agrimony* Agrimonia eupatoria, *water avens* Geum rivale, *and many others (see the list on page 140).*

You can also make variations of wetlands according to circumstances. These might include a wet ditch, fed either naturally or artificially, or a wet meadow if you happen to have an existing damp area in the garden. Both of these could add considerable interest to your garden, but if, by chance, you live in an area of high rainfall and happen to be well-supplied with acid water, you might be able to create a true bog-garden or, who knows, the very first wildlife garden 'raised-bog'!

Enjoying wildlife gardens

Apart from the obvious conservation function that wildlife gardens can have, as we have stressed throughout the book, they are also there to be enjoyed and appreciated like any other garden. As the garden that you have created develops, the more it will offer both to you and your visitors. When researching this book, we visited quite a number of wildlife garden owners, both in Britain and abroad, and it was manifestly obvious from talking to them just how much they enjoyed their gardens. It is true, there is a fair amount of work involved in creating and maintaining a good wildlife garden, especially a large one, but there should also be time to appreciate it, show it to other people and find out more about the visiting wildlife.

It is actually a good starting point, when planning or creating any wildlife garden, to go and visit other people who have tried to do something similar. Although wildlife gardening has only become popular in the last few years, it is surprising just how long a few people have been practising it, and they have built up considerable experience in what to do – and what not to bother with! It will also help you with your own ideas – mixtures of flowers that go well together, a flower you had not thought of, covered with butterflies or hoverflies, or a different design of bat-box. There are so many possibilities, and we are learning all the time how nature responds to gardening. The more ideas are shared and exchanged, the greater the success rate of wildlife gardening.

Visiting other wildlife gardens is not difficult, though well established ones are still relatively few and far between. In Britain, you can start with the excellent 'yellow book' published annually that lists private gardens open to the public in aid of charity – *Gardens of England and Wales; a complete list of private gardens open in aid of the National Gardens Scheme*. This lists all the relevant gardens by county, and usually gives enough information for you to judge whether wildlife gardening plays a significant part in the *raison d'etre* of each garden. We have discovered some real gems from this list, with marvellous mature wildlife garden features, though admittedly some that have sounded good have been disappointing. Although many are only open on specified days, most owners are only too pleased to let you come by private arrangement on another date if approached courteously, and will be grateful for your charitable contribution. Gardeners in general are a very helpful and friendly race, only too pleased to share information, plants or whatever!

Major horticultural shows nowadays often have one or two wildlife theme gardens. The Royal Horticultural

Society's main show at Chelsea in May each year has had several, as have the National or International Garden Festivals held regularly in Britain, Germany and elsewhere. All such demonstration gardens purvey ideas of what can be achieved, show new materials and, perhaps most important of all, inspire you. Admittedly, many of them will have been created within a matter of weeks, using more resources than you are likely to have at your disposal, but it does show how quickly results can be obtained. The most inspiring long term results come from countries such as West Germany, where there are some particularly effective wildlife gardens, and Holland, where the experience of

WILDLIFE GARDENS TO VISIT

The National Trust – both for Scotland, and England and Wales – has a number of gardens that have wildflower or wildlife areas within them, though this information is not always clearly described under each garden. You can usually find out more by approaching the relevant regional offices, if you can talk to the right person. There are also organisations which have set up wildlife gardens to demonstrate the principles to the public – for example, at the Centre for Alternative Technology in Wales or at visitor centres of various conservation organisations.

Wildlife garden at Careby Manor, Lincolnshire.

ecologically based wild gardening goes back many years; here there are several 'natural' parks, created using native flora, that are easily accessible to the visitor.

Your own garden, of course, will give you more pleasure than all these, because you are there for most of the time and because you created it, even though it may still be in its early stages. One of the nicest aspects of embarking on a project like this is the way in which you can observe things changing over the years; the pond that was just a brown sludgy hole becomes a living green gem in its first year, and then in the second

---

The presence of a pond in your garden allows you to occasionally bring pond life such as tadpoles into the house for closer study and enjoyment in an aquarium.

---

year dragonflies emerge from it and frogs spawn into it. Or your meadow gradually becomes more colourful as individual plants you put in spread or mature, and others appear. Or, the way in which your nectar border gradually becomes known to more and more of the neighbourhood hoverflies and butterflies. All these changes may be overlain by the climatic and natural changes that take place every year, and the normal fluctuations in numbers of plants and animals, but you will also be able to detect the seasonal and annual patterns in your own garden because you see it so often.

Take moths, as an example. You rarely see many of these insects except a few at the windows or fluttering about a light. However, we have several friends who run moth-traps in their gardens. Among these, it turns out that one has recorded an extraordinary total of about 440 moth species in his garden, and this is just the larger macro-moths, not the small micro-moths, of which there may be as many again. Another friend, after only two years' investigations, has recorded about 350 species of the larger moths in his garden, and he regularly watches half a dozen bats, of at least two

---

STUDY YOUR GARDEN

You will undoubtedly gain more pleasure from your garden if you study it closely. We are not suggesting that you peg out metre square plots to record the population dynamics of the weed seedlings, or mark the wings of every butterfly that visits your garden to see if it returns (though both these studies would yield fascinating results). You may feel tempted by such studies eventually, but at first it is fascinating to find out a little more about what is using your garden, and it is likely to prove very surprising. By taking two or three examples of studies made by professionals or dedicated amateurs in their gardens, you can begin to see the possibilities.

species, flying over his moth-trap. These traps attract the moths by light, so there is no doubt that individuals which otherwise might never have entered the garden will come to it, but it does give an idea of just how many species are around the average garden. It is also remarkable just how many individual moths there can be – one of these two gardeners has had an estimated 15,000 moths in one night! Unless you run a moth-trap, or even just venture into the garden at night, you remain in total ignorance of all this activity.

In another garden, a suburban one in Leicestershire, the owners (who are both professional entomologists) have made strenuous efforts to find out what has been visiting their garden. They have kept permanent traps in place, run moth-traps, netted butterflies and tried various other methods. Just to give two examples of what they have discovered: over a period of eight years they hand-netted 16,626 individual butterflies from 21 different species; while from another study, they have recorded 530 different species of Ichneumon wasp in the garden, including both new records for Britain and others new to science! We quote these examples, not in the expectation that other people will embark on similar studies, but just to show the incredible amount of life that there can be in ordinary gardens and how much of it passes unnoticed.

There are many books on how to study wildlife, and most of the ideas and techniques apply equally to gardens, so there is little need to go into them here. However, we will mention a few quick suggestions which can be particularly revealing and are well worth trying for yourself.

First, try going out at night into the garden with a powerful torch, at several different times of year and times of night. Just wander about checking flower-

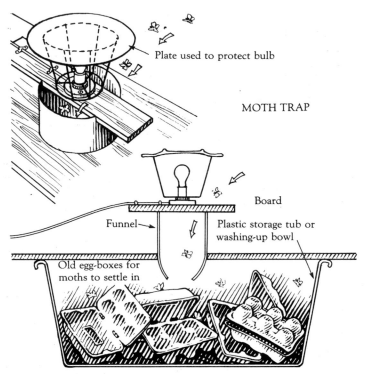

MOTH TRAP

Plate used to protect bulb

Board

Funnel

Plastic storage tub or washing-up bowl

Old egg-boxes for moths to settle in

A simple design of moth trap, that can be easily made at home using a plastic bowl, a funnel, some wood and a light will attract numerous species.

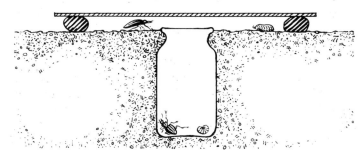

A simple trap for invertebrates can be made by hollowing out a potato, and half-burying it in the soil, with the entrance at ground level.

A pitfall trap looks like this in cross-section, and it simply collects a sample of the more mobile nocturnal creatures that you can examine each morning.

Even a simple moth-trap, based on a bare light and some cardboard egg-boxes will allow you to investigate what comes to your garden at night.

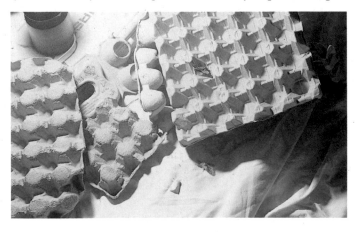

By encouraging more birds into your garden to feed, you have the opportunity of much closer views of wild birds.

heads, leaves, dark corners, the lawn or the pond with the torch to get an idea of the different range of life that wakes up at night. Not only moths, bats, owls and hedgehogs, but also spiders, butterfly or hawkmoth caterpillars, ground beetles, slugs and many more come alive at night. On other occasions, try walking slowly or sit quietly and just listen: you may hear the high-pitched 'pinging' of bats, the call of an owl, the rustle of voles or the shuffling of a hedgehog or even the call of a fox, if you are patient, quiet and warmly clothed. If you sit quietly somewhere with your torch covered either by red cellophane, or even brown paper, you may be lucky enough to get a glimpse of some of the mammals that visit your garden when you are not using it.

A moth-trap is well worth acquiring if you can, and you will be amazed not only by the variety and numbers of moths but also by their great beauty. However, on a much simpler, cheaper level, you can get a close view of some of the smaller creatures which run around on the ground at night by setting some pitfall traps. These are simply jars or yoghurt cartons, sunk into the soil so that their rims do not project above ground, then covered by a slightly raised tile or stone to allow the animals entry but preventing the traps from filling with rainwater. You can put them in different habitats, and either bait them or simply leave them as

unsigned holes in the road for unwary nocturnal travellers. If you examine them each morning, you will find ground beetles, spiders, woodlice, earwigs and various unexpected things – or the half-eaten remains of some – as indications of what has been moving around in your garden at night.

Secondly, take a closer look at your pond life. Just about everyone is familiar with the idea of pond-dipping, but there are a couple of other suggestions if you have a pond in the garden and want to study it a little more closely. The first is that you can try lying down next to the pond. Get really close to the water, and just watch what goes on below the surface and on and around the floating leaves for a quarter of an hour or so. You may not be able to identify the participants at first, but you will get a much clearer idea of what the various pond inhabitants are doing and how they inter-relate with one another. When you catch them in a jar, you interrupt their normal activities and discover little about them except their names; with a good clear water-level viewing patch, you can see what really goes on as dragonfly nymphs shoot out their masks, ostracods rocket past, while little picture-winged flies wave semaphor signals with their coloured wings at potential mates on nearby pond-weed leaves.

You can also set up a small temporary or permanent aquarium in the house containing various plants and animals from the pond. You can raise tadpoles or newt eggs, or just watch how some of the other creatures interact, with the even better view provided by the glass sides of the aquarium. There are plenty of books about maintaining aquaria, though one useful general rule is to keep it out of direct sun so that it does not overheat.

Thirdly, try going out into the garden round about dawn in May or June. This is the time of peak territorial activity for birds, and a quiet prowl, or even sitting in a shivering huddle somewhere, will give you a good idea of what is nesting in your garden. It is likely that all the singing males will have mates on eggs or young, so you can get a rough estimate of the number of nests around by looking for the individual males. You will not know exactly where each bird is nesting, and it may not be in your garden, but you will learn a little more of what is going on.

One final suggestion, to find out which insects are

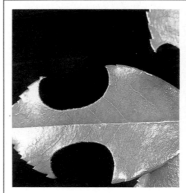

These neat holes in rose leaves are made by female leaf-cutter bees gathering nest material. Although considered unsightly by rose-growers, they do no harm, and are a sign of success that you are encouraging more insects to the garden. If you are lucky, you may see the female carrying off the cut leaf (right).

using your new habitats in daytime, is to go out with a sweep net on a warm summer's day. A sweep net is a large general insect net, made of tough material, that you use fairly vigorously to collect insects both from above the vegetation and within it. After several passes across and through the vegetation, fold the top over and examine your catch. At first, you may only see the

A sweep net, below, can be used to investigate the fascinating insect life of a flowery meadow.

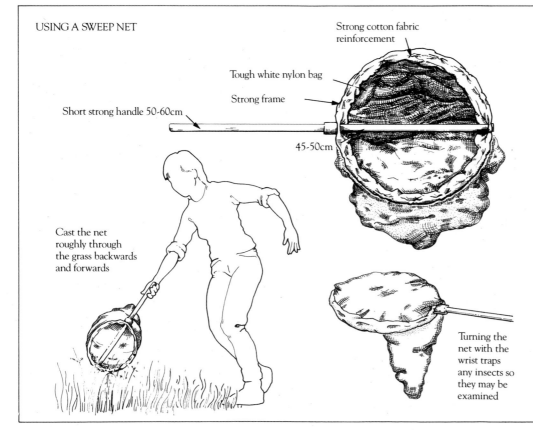

USING A SWEEP NET

Strong cotton fabric reinforcement

Tough white nylon bag

Strong frame

Short strong handle 50-60cm

45-50cm

Cast the net roughly through the grass backwards and forwards

Turning the net with the wrist traps any insects so they may be examined

The best way of sampling the insects found in and around your garden is to construct, or buy, a sweep net.

The net should be of closely woven, preferably white, nylon, reasonably tough, and with a depth twice the diameter of the rim so the net can be turned at the end of each stroke to seal any captured insects in the bag. The net should be connected to the heavy gauge wire rim by some cotton fabric wrapped over the wire. A cross-member will add rigidity. The net should be connected to a short, sturdy handle.

Cast it through the grass, backwards and forwards, dipping just below the vegetation level so that the net actually gets in amongst the grass.

A special platform, built out over the fringing vegetation, provides an excellent means of studying pond life at close quarters without getting wet.

larger insects, trying to flutter their way out, but after a while you begin to see the green capsid bugs, the sawflies, the tiny parasitic wasps with long antennae, the colourful flower beetles, the young stages of a grasshopper or cricket and many more. You may not have discovered much about what they were doing, but it can be very illuminating just to see what is there, even in quite a new habitat.

WILDLIFE NOTEBOOK

Do make notes on your wildlife garden – the changes, any new species or acquisitions, observations on how the plants and animals are doing – and make sure they are clearly dated. Photographs, too, can provide an interesting visual record of the development and change in your garden. Perhaps in 10 years time, you will look back at your first wildlife garden notes, and be surprised – and rewarded – by the sudden realisation of all that you have started.

USEFUL ORGANISATIONS

British Butterfly Conservation Society
Tudor House,
Quorn,
Loughborough, Leicestershire LE12 8AD

British Hedgehog Preservation Society
Knowbury House,
Knowbury,
Ludlow,
Shropshire

British Trust for Conservation Volunteers (BTCV)
London Office: 80, York Way,
London N1 9AG
Head Office: 36, St. Mary's Street,
Wallingford,
Oxfordshire OX10 0EH

British Trust for Ornithology (BTO)
Beech Grove,
Tring,
Hertfordshire HP23 5NR

Centre for Alternative Technology
Machynlleth,
Powys,
Wales SY20 9AZ

The Cottage Garden Society
c/o Mrs Philippa Carr
15 Faenol Avenue
Abergele
Clwyd LL22 7HT

Fauna and Flora Preservation Society (FFPS)
Zoological Gardens,
Regent's Park,
London NW1 4RY

Field Studies Council
62 Wilson Street,
London EC2A 2BU

Friends of the Earth (FOE)
26–28 Underwood Street,
London N1 7JQ

Greenpeace
36 Graham Street,
London W1 2JX

Henry Doubleday Research Association
Convent Lane,
Bocking,
Braintree,
Essex

Nature Conservancy Council (NCC)
Northminster House,
Peterborough,
PE1 1UA

Royal Horticultural Society
Vincent Square,
London SW1P 2PE

Royal Society for Nature Conservation (RSNC)
The Green,
Nettleham,
Lincoln LN2 2NR

Royal Society for the Protection of Birds (RSPB)
The Lodge,
Sandy,
Bedfordshire SG19 2DL

Seed Bank
44 Albion Road,
Sutton,
Surrey SM2 5TF

Trust for Urban Ecology (TRUE)
South Bank House,
Black Prince Road,
London SE1

WATCH
22, The Green,
Nettleham,
Lincoln LN2 2NR

Young Ornithologists' Club (YOC)
The Lodge,
Sandy,
Bedfordshire SG19 2DL

SUPPLIERS

NATIVE PLANTS AND SEEDS

Most native trees and shrubs should be widely available (*although not always of native origin*) so the sources listed below are mainly those supplying seeds and/or herbaceous material. Please send stamped addressed envelope with any enquiry, and make sure that the nursery is open to the public before visiting.

Careby Manor Gardens (*plants*)
Careby,
Stamford,
Lincolnshire PE9 4EA

John Chambers (*seeds, plants*)
15 Westleigh Road,
Barton Seagrave,
Kettering,
Northamptonshire NN15 5AJ

Emorsgate Seeds (*seeds*)
Emorsgate,
Terrinton St. Clement,
King's Lynn,
Norfolk PE34 4NY

Kingsfield Tree Nursery (*plants, trees and shrubs*)
G, J E Peacock,
Broadenham Lane,
Winsham,
Chard,
Somerset TA20 4JF

Landlife Wild Flowers Ltd (R P A) (*seeds, plants*)
The Old Police Station,
Lark Lane,
Liverpool L17 8UU

Manor House Herbs (*plants*)
Wadeford,
Chard,
Somerset TA20 3AO

Naturescape (*seeds, plants*)
Little Orchard,
Main Street,
Whatton in the Vale,
Nottinghamshire NG13 9EP

Nordybank Nurseries (*seeds, plants*)
Clee St Margaret,
Craven Arms,
Shropshire SY7 9EF

Oak Cottage Herb Farm (*plants*)
Ruth Thompson
Nesscliffe,
Shropshire DY4 1DB

Poyntzfield Herb Nursery (*plants, medicinal*)
Conon Bridge,
Black Isle,
Ross-shire,
Highland

The Seed Bank (*seeds, seed exchange*)
Cowcombe Farm,
Gipsy Lane,
Chalford,
Stroud,
Gloucestershire GL6 8HP

Suffolk Herbs (*seeds, plants*)
Sawyers Farm,
Little Cornard,
Sudbury
Suffolk CO10 0NY

Yew Tree Herbs (*plants*)
Holt Street,
Nonington,
Dover,
Kent CT15 4JS

PONDS AND WATER GARDENS

Highlands Water Gardens
Solesbridge Lane,
Chorley Wood,
Hertfordshire WD3 5SZ

Stapeley Water Gardens Ltd
Stapeley,
Nantwich,
Cheshire CW5 7LH

Maydencroft Aquatic Nurseries
Maydencroft Lane,
Gosmore,
Hitchin,
Hertfordshire

G Mimack
The Water Plant Farm,
Woodholme Nursery,
Stock,
Essex

Wildwoods Water Garden Centre
Theobalds Park Road,
Crews Hill,
Enfield,
Middlesex EN2 9BP

Rawmat Bentonite Clay
Rawell Marketing Ltd,
Carr Lane,
Hoylake,
Merseyside L47 4AX

RECOMMENDED FURTHER READING

Baines, C. 1985.
How to make a wildlife garden.
Elm tree books, London.

Brookes, J. and Beckett, K.A. 1987.
Gardeners' Index of Plants and Flowers.
Dorling Kindersley, London.

BTCV. Wildlife Gardening Resource Pack.
British Trust for Conservation Volunteers, London.

Buczacki, S. 1986.
Ground Rules for Gardeners.
Collins, London.

Chinery, M. 1986.
The Living Garden.
Dorling Kindersley, London.
1979.
The Family Naturalist.
Roxby Press Productions.
1986.
Insects of Britain and Western Europe.
Collins, London.

Colebourn, P. and Gibbons, R. 1987.
Britain's Natural Heritage.
Blandford Press, Poole.

Fitter, R., Fitter, A. and Blamey, M. 1985.
Wild Flowers of Britain and Northern Europe.
4th ed. Collins, London.

Fitter, R. and Manuel, R. 1986.
Field Guide to Freshwater Life of Britain and North-West Europe.
Collins, London.

Gibbons, R. 1986.
Dragonflies and Damselflies of Britain and Northern Europe.
Country Life Books, London.

Hamilton, G. 1987.
Successful Organic Gardening.
Dorling Kindersley, London.

International Bee Research Association. 1981.
Garden Plants Valuable to Bees.
IBRA, London.

McEwan, H. 1982.
Seed Growers Guide to Herbs and Wildflowers.
Suffolk Herbs, Suffolk.

Oates, M. 1985.
Garden Plants for Butterflies.
Brian Masterton & Ass., Fareham.

Owen, J. 1983.
Garden Life.
Chatto & Windus, London.

Richardson, P. 1985.
Bats.
Whittet books, London.
Also see other Whittet books on **Squirrels, Foxes, Hedgehogs, Robins, Garden Creepy-Crawlies.**

Rothschild, M. and Farrell, C. 1983.
The Butterfly Gardener.
Michael Joseph/Rainbird, London.

Soper, T. 1985.
Discovering Animals.
BBC Books, London.

Stevens, D. 1984.
Making a Garden.
World's Work Ltd, Surrey.

Stevens, J. 1987.
National Trust Book of Wildflower Gardening.
Dorling Kindersley, London.

Wright, M. 1984.
The Complete Handbook of Garden Plants.
Michael Joseph, London.

Wright, T. 1985.
Labour-saving Gardening.
Penguin, Middlesex.

PROTECTED PLANTS
in the
UNITED KINGDOM

Although wildlife gardens are immensely valuable in augmenting threatened wildlife habitats, they can never replace the natural associations of plants and animals which have taken hundreds, or even thousands, of years to establish. It follows that even though a reasonably effective imitation can be achieved, this should never be at the expense of the natural habitat. This is why plants should never *be dug up from the wild, and the wildlife gardener must accept some responsibility in trying to make sure that any purchases made of plants or seed are not from a wild source. We would not recommend collecting seed from the wild for the same reasons, even if the plant is not a protected species. Similarly, if you are making a rock garden, guard against buying weathered limestone, as it may well come from limestone pavement, one of our rarest and strangest natural habitats. The following plants are protected under the terms of the* **Wildlife and Countryside Act 1981***. They must not be uprooted or picked, nor can seed be collected from them. (No wild plants can be uprooted except in special circumstances.)*

Adder's-tongue spearwort *Ranunculus ophioglossifolius*
Alpine catchfly *Lychnis alpina*
Alpine sow-thistle *Cicerbita alpina*
Blue heath *Phyllodoce caerulea*
Branched horsetail *Equisetum ramosissimum*
Bedstraw broomrape *Orobranche caryophyllacea*
Ox-tonge broomrape *Orobranche loricata*
Thistle broomrape *Orobranche reticulata*
Brown galingale *Cyperus fuscus*
Cambridge milk-parsley *Selinum carvifolia*
Creeping marshwort *Apium repens*
Jersey cud-weed *Gnaphalium luteo-album*
Red-tipped cud-weed *Filago lutescens*
Diapensia *Diapensia laponica*
Dickie's bladder-fern *Cystopteris dickieana*
Early star-of-Bethlehem *Gagea bohemica*
Fen ragwort *Senecio paludosus*
Fen violet *Viola persicifolia*
Field cow-wheat *Melampyrum arvense*
Field eryngo *Eryngium campestre*
Field wormwood *Artemisia campestris*
Alpine fleabane *Erigeron borealis*
Small fleabane *Pulicaria vulgaris*
Foxtail stonewort *amprothamnium papulosum*
Alpine gentian *Gentiana nivalis*
Fringed gentian *Gentianella ciliata*
Spring gentian *Gentiana verna*

Greater yellow-rattle *Rhinanthus serotinus*
Grass-polly *Lythrum hyssopifolia*
Cut-leaved germander *Teucrium botrys*
Water germander *Teucrium scordium*
Green hound's-tongue *Cynoglossum germanicum*
Sickle-leaved hare's-ear *Bupleurum falcatum*
Small hare's-ear *Bupleurum baldense*
Holly-leaved naiad *Najas marina*
Red helleborine *Cephalanthera rubra*
Young's helleborine *Epipactis youngiana*
Killarney fern *Trichomanes speciosum*
Lady's-slipper *Cypripedium calceolus*
Least adder's-tongue *Ophioglossum lusitanicum*
Least lettuce *Lactuca saligna*
Lundy cabbage *Rhynchosinapis wrightii*
Martin's ramping-fumitory *Fumaria martinii*
Fen orchid *Liparis loeselii*
Ghost orchid *Epipogium aphyllum*
Lizard orchid *Himantoglossum hircinum*
Military orchid *Orchis militaris*
Monkey orchid *Orchis simia*
Pennyroyal *Mentha pulegium*
Perennial knawel *Scleranthus perennis*
Pigmyweed *Crassula aquatica*
Cheddar pink *Dianthus gratianopolitanus*
Chidling pink *Petrorhagia nanteuilii*
Plymouth pear *Pyrus cordata*
Purple colt's-foot *Homogyne alpina*

Purple spurge *Euphorbia peplis*
Ribbon-leaved water-plantain *Alisma gramineum*
Rock cinquefoil *Potentilla rupestris*
Alpine rock-cress *Arabis alpina*
Bristol rock-cress *Arabis stricta*
Rough marsh-mallow *Althaea hirsuta*
Sand crocus *Romulea columnae*
Norwegian sandwort *Arenaria norvegica*
Teesdale sandwort *Minuartia stricta*
Drooping saxifrage *Saxifraga cernua*
Tufted saxifrage *Saxifraga caespitosa*
Sea knotgrass *Polygonum maritimum*
Rock sea-lavender *Limonium paradoxum*
Rock sea-lavender *Limonium recurvum*
Round-headed leek *Allium sphaerocephalon*
Slender cottongrass *Eriphorum gracile*
Small alison *Alyssum alysoides*
Small restharrow *Ononis reclinata*
Snowdon lily *Lloydia serotina*

Fingered speedwell *Veronica triphyllos*
Spiked speedwell *Veronica spicata*
Early spider-orchid *Ophrys sphegodes*
Late spider-orchid *Ophrys fuciflora*
Star fruit *Damasonium alisma*
Starved wood-sedge *Carex depauperata*
Stinking goosefoot *Chenopodium vulvaria*
Stinking hawk's-beard *Crepis foetida*
Triangular club-rush *Scirpus triquetrus*
Viper's-grass *Scorzonera humilis*
Whorled Solomon's-seal *Polygonatum verticillatum*
Wild cotoneaster *Cotoneaster integerrimus*
Wild gladiolus *Gladiolus illyricus*
Wood calamint *Calamintha sylvatica*
Alpina woodsia *Woodsia alpina*
Oblong woodsia *Woodsia ilvensis*
Downy woodwort *Stachys germanica*
Limestone woundwort *Stachys alpina*

ACKNOWLEDGMENTS

The authors are very grateful to the owners of various wildlife gardens in Britain and Germany, and especially to the Callans at Winllan in Wales. Also, grateful thanks are due to Matthew Oates for many helpful discussions on butterflies in the garden. Our thanks also go to Charles Stitt for all his efforts in getting the artwork right under difficult circumstances. And finally thanks to Robin and Kit who have put up with both parents writing at once and months of garden visits!

Artwork by CHARLES STITT based on original drawings by Liz Gibbons, with the exception of page 148 by JAMES HUGHES and page 122 by James Hughes based on original drawings by Liz Gibbons.

All Photographs by LIZ AND BOB GIBBONS/NATURAL IMAGE with the exception of:

AQUILA PHOTOGRAPHICS Michael Leach 6, 50–51, Robert Maeir 72–73; MICHAEL BOYS 11; HAMLYN PUBLISHING GROUP/PETER LOUGHRAN 95; NATURAL IMAGE Ian Callan/Bob Gibbons 42; Robert Dickson 74, 84, 106; Robin Fletcher 35, 127, 143 bottom; Alec Harmer 12 bottom; Andrew Lawson 88–89; Julie Meeke 85; Peter Wilson 16, 21 bottom left, 22 top, 24–25, 52, 61, 63 right, 67, 68, 79, 81, 87, 148 right.

INDEX

Figures in italics refer to captions